If You're Trying to
Get Better Grades
&
Higher Test Scores
in Math,

You've Gotta Have This Book !

Grades 6 & Up

by Imogene Forte
& Marjorie Frank

Incentive Publications, Inc.
Nashville, Tennessee

Illustrated by Kathleen Bullock
Cover by Geoffrey Brittingham
Edited by Charlotte Bosarge

ISBN 978-0-86530-576-2

9   10   10

Printed by Sheridan Books, Inc., Chelsea, Michigan • July 2010
www.incentivepublications.com

Contents

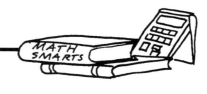

GET SHARP . . . on Number Concepts & Relationships   55

GET SHARP . . . on Operations with Whole Numbers   75

GET SHARP . . . on Statistics, Graphing & Probability   153

GET SHARP . . . on Pre-Algebra   173

GET SHARP . . . on Problem Solving 191

GET SHARP . . . on Math Terms 209

INDEX 235

─── Get Ready ───

Get ready to get smarter. Get ready to be a better student and get the grades you are capable of getting. Get ready to feel better about yourself as a student. Lots of students would like to do better in school, and lots of their parents and teachers would like them to as well! Lots of students CAN do better. But it doesn't happen overnight, and it doesn't happen without some thinking and trying. So are you ready to put some energy into getting more out of your learning efforts? Good! The first part of getting ready is *wanting* to do better—motivating yourself to get moving on this project of showing how smart you really are. The **Get Ready** part of this book will help you do just that: get inspired and motivated. It also gives you some wonderful and downright practical ways to organize yourself, your space, your time, and your homework. Even more than that, it gives you tips to use right away to make big improvements in your study habits.

─── Get Set ───

Once you've taken a good, hard look at your goals, organization, and study habits, you can move on to other skills and habits that will get you set up for more successful learning. The **Get Set** part of this book gives you ready-to-use tools for sharpening thinking skills and helping you get the most out of your brain. Then, it adds a quick and effective crash-course on finding information in the library, on the Internet, and from many other sources. Top this off with a great review of tools and skills you need for good studying. It's all right here at your fingertips—how to read carefully, listen well, summarize, outline, take notes, create reports, study for tests, and take tests. Take this section seriously, and you're bound to start making improvements immediately.

─── Get Sharp ───

Now you're ready to mix those good study habits and skills with the content that you want to learn. The **Get Sharp** sections of this book contain all kinds of facts and explanations, processes and definitions, lists, and how-to information. These sections cover all of the basic areas of math that you study in school. They are loaded with the information that you need to do your math homework. You will find this part of the book to be a great reference tool PLUS a *How-To Manual* for many topics and assignments. Keep it handy whenever you do an assignment in math.

How to Use This Book

Students

Students—this can be the ultimate homework helper for your math assignments and preparation. Use the *Get Ready* and the *Get Set* sections to strengthen your general preparation for study and sharpen your study skills. Then, have the book nearby at all times when you have math work to do at home, and use the *Get Sharp* sections to . . .

    . . . reinforce a topic you've already learned.

    . . . get fresh and different examples of something you've studied.

    . . . check up on a math fact or definition.

    . . . get a quick answer to a math question.

    . . . get clear on something you thought you knew but now aren't sure about.

    . . . guide you in problem-solving processes.

    . . . check yourself to see if you've got a fact or process right.

    . . . review a topic in preparation for a test.

Teachers

This book can serve multiple purposes in the classroom. Use it as . . .

    . . . a reference manual for students to consult during learning activities or assignments.

    . . . a reference manual for you to consult on particular rules, terms, forms, and skills.

    . . . an instructional handbook for particular math topics.

    . . . a remedial tool for individuals or groups who need a review of a particular math topic.

    . . . a source of advice for parents and students regarding homework habits.

    . . . an assessment guide to help you gauge student mastery of math processes or skills.

    . . . a source of good resources for making bridges between home and school.
        (For starters, send a copy of the letter on page 17 home to each parent. Use any other pages, particularly those in the "Get Ready" and "Get Set" sections, as send-home pieces.)

Parents

The *Get Ready* and *Get Set* sections of this book will help you to help your child improve study habits and sharpen study skills. It can serve as a motivator and a guide, and take the burden off you! Then, use the *Get Sharp* sections as a knowledge and process back-up guide for yourself.

It's a handbook you can consult to. . .

    . . . refresh your memory about a math process, term, rule, or fact.

    . . . clear up confusion about math rules, skills, and other questions.

    . . . provide useful homework help to your child.

    . . . reinforce the good learning your child is doing in school.

    . . . gain confidence that your child is doing the homework right.

GET READY →

Get Motivated

Dear Student,

Nobody can make you a better student. Nobody can even make you WANT to be a better student. But you CAN be. It's a rare kid who doesn't have some ability to learn more, do better with assignments and tests, feel more confident as a student, or get better grades. YOU CAN DO THIS! You are the one (the only one) that can get yourself motivated.

Probably, the first question is this: "WHY would you want be a better student?" If you don't have an answer to this, your chances of improving are not so hot. If you do have answers, but they're like the ones on pg. 15, your chances of improving still might be slim. THAT student figured it out, and decided that these are NOT what really motivate him. Now, we don't mean to tell you that it's a bad idea to get a good report card, or get on the honor roll, or please your parents. We're not trying to say that getting into college is a poor goal or that there's anything wrong with getting ready for high school either.

But—if you are trying to motivate yourself to be a better student, the reasons need to be about YOU. The goals need to be YOUR goals for your life right now. In fact, if you are having a hard time getting motivated, maybe it is just BECAUSE you're used to hearing a lot of "shoulds" about what other people want you to be. Or maybe it's because the goals are so far off in some hazy distant future that it's impossible to stay focused on them.

So it's back to the question, "Why try to be a better student?" Consider these as possible reasons why:

- to make use of your good mind (and NOT short-change yourself by cheating yourself out of something you could learn to do or understand)

- to get involved—to change learning into something YOU DO instead of something that someone else is trying to do TO you

- to take charge and get where YOU WANT TO GO (It's YOUR life, after all)

- to learn all you can for YOURSELF—because the more you know, the more you think, and the more you understand—the more possibilities you have for what you can do or be in your life RIGHT NOW and in the future

Follow the "Get Motivated Tips" on the next page as you think about this question. Then write down a few reasons of your own to inspire you toward putting your brain to work, showing how smart you are, and getting even smarter.

Sincerely,

Imogene and Marjorie

Better Grades & Higher Test Scores / MATH
©Incentive Publications, Inc., Nashville, TN

Why should I be a better student ?

To please my parents
To please my teachers
To impress other kids
To impress my parents' friends
So people will like me better
To keep from embarrassing my parents
To do as well as my older brother
To do better than my sister
So teachers treat me better
To do as well as my mom or dad did in school
To get the money my parents offer for good grades
To get the privileges my parents offer for good grades
To get well-prepared for high school
To make a lot of money when I finish school
To get a good report card
To get into college

None of these really motivate me much at all.

Get Motivated Tips

1. Think about why you'd want to do better as a student.

2. Think about what you'd gain now from doing better.

3. Get clear enough on your motivations to write them down.

4. Set some short-term goals *(something you can improve in a few weeks).*

5. Think about what gets in the way of doing your best as a student.

6. Figure out a way to change or eliminate something that interferes.

(Use the form on page 16 to record your thoughts and goals.)

Get Ready Tip # 1

Set realistic goals. Choose something you actually believe you can do. And, you'll have better chance of success if you set a short time frame for a goal.

1._____

_____

2._____

_____

3._____

_____

Why do I want to be a better student? What difference would it make for me, now and in the future?
(Write a few reasons.)

What changes could I make in the near future?
(Write two short-term goals—things that, realistically, you could improve in the next month.)

1._____

2._____

What gets in the way of good grades or good studying for me?
(Name the things, conditions, or distractions that **most often** keep you from doing your best as a student.)

1._____

2._____

3._____

4._____

What distraction am I willing to eliminate?
(Choose one of the interferences above that you'd be willing to try changing or getting rid of for the next month.)

1._____

16

Better Grades & Higher Test Scores / MATH
©Incentive Publications, Inc., Nashville, TN

Dear Parent:

What parent doesn't want his/her child to be a good student? Probably not many! But how can you help yours get motivated to do the work it takes? You can't do it for her (or him), but here are some ideas to help students as they find it within themselves to get set to be good students:

Read the letter to students (page 14). Help your son or daughter think about where she/he wants to go, what reasons make sense to her or him for getting better grades, and what benefits he/she would gain from better performance as a student.

Help your child make use of the advice on study habits. (See pages 18-28.) Reinforce the ideas, particularly those of keeping up with assignments, going to class, and turning in work on time.

Provide your child with a quiet, comfortable, well-lit place that is available consistently for study. Also provide a place to keep materials, post reminders, and display schedules.

Set family routines and schedules that allow for good blocks of study time, adequate rest, relaxing breaks, and healthy eating. Include some time to get things ready for the next school day and some ways for students to be reminded about upcoming assignments or due dates.

Demonstrate that you value learning in your household. Read. Learn new things. Show excitement about learning something new yourself. Share this with your kids.

Keep distractions to a minimum. You may not be able to control the motivations and goals of your child, but you can control the telephone, computer, Internet, and TV. These things actually have on-off switches. Use them. Set rules and schedules.

Help your child gather resources for studying, projects, papers, and reports. Try to be available to take her or him to the library, and offer help tracking down a variety of sources. Try to provide standard resources in the home (dictionaries, thesaurus, computer encyclopedia, etc.).

DO help your student with homework. This means helping straighten out confusion about a topic (when you can), getting an assignment clear, discussing a concept or skill, and perhaps working through a few examples along with the student to make sure he/she is doing it right. This kind of involvement gives a chance to extend or clarify the teaching done in the classroom. Remember that the end goal is for the student to learn. Don't be so insistent on the student "doing it himself" that you miss a good teaching or learning opportunity.

Be alert for problems, and act early. Keep contact with teachers and don't be afraid to call on them if you see any signs of slipping, confusion, or disinterest on the part of your child. It is easier to reclaim lost ground if you catch it early.

Try to keep the focus on the student's taking charge of meeting his/her own goals rather than on making you happy. This can help get you out of a nagging role and get some of the power in the hands of the student. Both of these will make for a more trusting, less hostile relationship with your child on the subject of school work. Such a relationship will go a long way toward supporting your child's self motivation to be a better student.

Sincerely,

Imogene and Marjorie

Get Organized

9:00 p.m. Izzy has a project on symmetry due tomorrow. She hasn't put it together yet.

Isabel! Get off the phone!

Goodbye, Jess. My math project is due tomorrow.

She clears a space in the clutter and starts looking for the folder where she put all her notes and drawings.

The markers she needs for her symmetry drawings are all dried up. She can't find her ruler.

Could I use a nail file as a straight edge?

She took some great photographs of real-life symmetry examples, but she forgot to get the film developed.

A terrible development! I never got this developed!

She may know a lot about quasars.
Maybe she's done some good research.
But she's not in a good place to show what she's learned because she is so disorganized. Don't repeat her mistakes.

Get Your Space Organized

Find a good place to study. Choose a place that . . .

. . . is always available to you.

. . . is comfortable.

. . . is quiet and as private as possible.

. . . has good lighting.

. . . is relatively uncluttered.

. . . is relatively free of distractions.

. . . has a flat surface large enough to spread out materials.

. . . has a place to keep supplies handy. (See page 19 for suggested supplies.)

. . . has some wall space or bulletin board space for posting schedules and reminders.

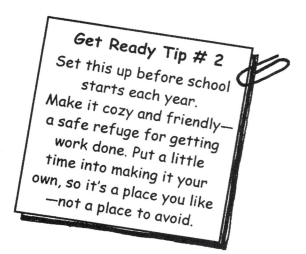

Get Ready Tip # 2
Set this up before school starts each year.
Make it cozy and friendly—a safe refuge for getting work done. Put a little time into making it your own, so it's a place you like —not a place to avoid.

Get Your Stuff Organized

Gather things that you will need for studying or for projects, papers, and other assignments. Keep them organized in one place, so you won't have to waste time running around looking. Here are some suggestions:

Also have:
an assignment notebook (See page 22)
a notebook for every subject
a book bag or pack to carry things back and forth
a schedule for your week (or longer)

Get set with a place to keep supplies.
(a bookshelf, a file box, a paper tray, a drawer, a plastic dish pan, a plastic bucket, a carton, or plastic crate)
Keep everything in this place at all times.
Return things to it after you use them.

Get Ready Tip # 3
Have a place to put things you bring home from school. This might be a shelf, a box, or even a laundry basket. Put your school things in there every time you come in the door—so important stuff doesn't get lost in the house or moved or used by other family members.

Supplies to Have Handy
a good light
a clock or timer
bulletin board or wall
(for schedule & reminders)
pencils, pens, erasable pens
erasers
colored pencils or crayons
markers
highlighters
notebook paper
scratch paper
drawing paper
typing-computer paper
index cards
sticky notes
poster board
folders
ruler, compass
tape, scissors
calculator
glue, rubber cement
paper clips, push pins
stapler, staples
standard references:
dictionary
thesaurus
current almanac
world maps
language handbook
writer's guide
encyclopedia (set or CD)
homework hotline numbers
homework help websites

Better Grades & Higher Test Scores / MATH
©Incentive Publications, Inc., Nashville, TN

Get Your Time Organized

It might be easy to organize your study and space and supplies, but it is probably not quite as easy to organize your time. This takes some work. First, you have to understand how you use your time now. Then you'll need to figure out a way to make better use of your time. Here's a plan you can follow right away to help you get your time organized:

Think about how you use your time now.

1. For one week, stop at the end of each day, think back over the day, and write down what you did in each hour-long period of time for the whole day.

2. Then look at the record you've kept to see how you used your time.

   Ask yourself these questions:
   Did I have any clear schedule?
   Did I have any goals for when
   I would get certain things done?
   Did I ever think ahead about
   how I would use my time?
   How did I decide what to do first?
   Did I have a plan or did I just get things done in haphazard order?
   Did I get everything done or did I run out of time?
   How much time did I waste?

8:00 am I left my homework folder on the breakfast table.

10:30 am I crammed for my math test between history and music class.

3:30 pm I talked on the phone until dinner.

5:00 pm I read magazines to warm up my brain for homework.

7:00-8:00 pm I couldn't miss my favorite TV show.

9:00 pm I started my math homework.

9:10 pm I fell asleep on my math book.

3. Next, start fresh for the upcoming week. Make a plan. Include:
   time that will be spent at school
   after-school activities
   meals
   study time
   family activities
   fun, sports, or recreational activities
   social activities or special events
   sleep time

Get Ready Tip # 4

When you plan your week's schedule, don't make it too tight or too rigid. Leave room for unexpected events.

4. Make sure you have an assignment notebook. When you plan your weekly schedule, transfer assignments from that notebook into your study time. *(Did you leave enough time to do all these assignments?)* Keep a copy of your calendar at home *and* in a notebook that you carry to school.

5. Make a Daily *To-Do* List *(For each day, write the things that must be done by the end of that day.)*

M 5th	T 6th	W 7th	TH 8th	F 9th
8 am–3 pm **School** Due: Science project Math: pg 115 Health: pg 78	Math test Due: English: 3 short stories History: Read Ch 7	Grammar quiz Due: French essay Math: pg 119	Due: History timeline Health: Ch 9 Math: pg 121	Due: English: autobiography English: Ch 4 poetry
4-7 pm volleyball practice 3:30 relax dinner	volleyball game 4:00 relax dinner	volleyball practice 3:30 relax dinner	volleyball practice 3:30 relax dinner	volleyball game 4 pm dinner
7-10 pm **Study Time** Math Test English: Read short stories Grammar quiz Start French essay History: finish Ch 7	**Study Time** Math: pg 119 Grammar quiz finish French essay Health: Ch 9 9:00 TV show	**Study Time** Math: p 121 Health: Ch 9 review quest. finish History timeline English: work on autobiography	7 pm Choir **Study Time** English: read Ch 4 English: finish autobiography	Bob's birthday party 7 pm

VB game 10 am Sat, football game Sat night, get supplies for health project
7 - 9 pm Sun Study Time: finish poetry chapter, start Health project

Long-Range Assignments (due next week): Health project—disease prevention
finish novel, English—report on novel

Get Ready Tip # 5

At all times—keep a copy of class outlines, schedules, or long-range class assignments at home.

Mon. TO DO List

study for Math test
finish 3 English short stories
review for Tues Grammar quiz
call Jon to schedule weekend study session
wash volleyball uniform
return library books
start French essay
check Internet for History info on Revolutionary War

Wed. TO DO List

finish History timeline
read Health, Ch 8
Math problems, pg 121
work on autobiography
shop for Bob's present
start English, Ch 4

Date	Subject	Assignment	Due Date
1/22	English	choose novel, write report	2/16
1/23	French	essay about family	2/7
1/25	English	3 short stories in book (pgs 35-55)	2/6
1/30	History	Revolutionary War timeline	2/8
1/30	Science	project on cells	2/5
1/31	English	Grammar quiz, Ch 7	2/7
1/31		autobiography	2/9
2/1	Math	read Ch 4 poetry	2/9
2/1	Math	unit 6 test	2/6
2/5	Math	problems, page 119	2/7
2/7	Health	problems, page 121	2/8
2/5		read Ch9	2/8

Get Your Assignments Organized

You can't do a very good job of an assignment if you don't have a clue about what it is. You can't possibly do the assignment well if you don't understand the things you're studying. So, if you want to get smarter, get clear and organized about assignments. It takes 7 simple steps to do this:

1. Listen to the assignment.

2. Write it down in an assignment notebook.
   (Make sure you write down the due date.)

3. If you don't understand the assignment—ASK.
   (Do not leave the classroom without knowing what it is you are supposed to do.)

4. If you don't understand the material well enough to do the assignment—
   TALK to the teacher. *(Tell him or her that you need help getting it clear.)*

5. Take the assignment book home every day.

6. Transfer assignments to your weekly or monthly schedule at home.

7. Look at your assignment book every day.

Get Yourself Organized

Okay, so your schedule is on the wall—all neat and clear. Your study space is organized. Your study supplies are organized. You have written down all your assignments, and you've got all your lists made. Great! But do you feel rushed, frenzied, or hassled? Take some time to think about the behaviors that will help YOU feel as organized as your stuff and your schedule.

Before you leave school . . .

STOP—take a few calm, unrushed minutes to think about what books and supplies you will need at home for studying. ALWAYS take the assignment notebook home.

When you get home . . .

FIRST—put your school bag in the same spot every day, out of the way of the bustle of your family's activities.

STOP—after relaxing, or after dinner, take a few calm, unrushed minutes to look over your schedule and review what needs to be done. Review your list for the day. Plan your evening study time and set priorities. Don't wait until it is late or you are very tired.

Before you go to bed . . .

STOP—take a few calm, unrushed minutes to look over the assignment notebook and the to-do list for the next day one more time. Make sure everything is completed.

THEN—put everything you need for the next day IN the book bag. Don't wait until morning. Make sure you have all the right books and notebooks in the bag. Make sure your finished work is all in the bag. Also, pack other stuff (for gym, sports, etc) at the same time. Put everything in one consistent place, so you don't have to rush around looking for it.

In the morning . . .

STOP—take a few calm, unrushed minutes to think and review the day one more time.

THEN—eat a good breakfast.

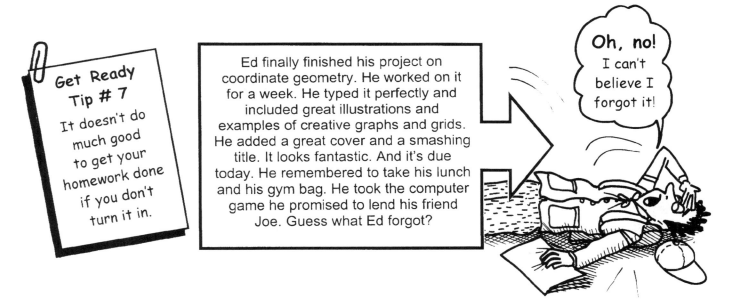

Get Ready Tip # 7

It doesn't do much good to get your homework done if you don't turn it in.

Ed finally finished his project on coordinate geometry. He worked on it for a week. He typed it perfectly and included great illustrations and examples of creative graphs and grids. He added a great cover and a smashing title. It looks fantastic. And it's due today. He remembered to take his lunch and his gym bag. He took the computer game he promised to lend his friend Joe. Guess what Ed forgot?

Oh, no! I can't believe I forgot it!

Get Healthy

If you are sick, or tired, or droopy, or angry, or nervous, or weak, or miserable, it is very hard to be a good student. It is hard to even use or show what you already know. Your physical and mental health is a basic MUST for doing as well as you would like to in school. So, don't ignore your health. Pay attention to how you feel. No one else can do that for you.

Get plenty of rest

If you're tired, nothing else works very well in your life. You can't think, concentrate, or pay attention, learn, remember, or study. Try to get 7 or 8 hours of sleep every night. Get plenty of rest on weekends. If you have a long evening of study ahead, take a short nap after school.

Eat well

You can't learn or function well on an empty stomach. And all that junk food (soda, sweets, chips, snacks) actually will make you more tired. Plus, it crowds the healthy foods out of your diet—the foods your brain needs to think well and your body needs to get through the day with energy. So eat a balanced diet, with lean meat, whole grains, vegetables, fruit and dairy products. Oh, and drink a lot of water—8 glasses a day is good.

Exercise

Everything in your body works better when your body gets a chance to move. Make sure your life does not get too sedentary. Do something every day to get exercise—walk, play a sport, play a game, or run. It's a good idea to get some exercise before you sit down to study, too. Exercise helps you relax, unwind, and de-stress. It's good for stimulating your brain.

Relax

Your body and your mind need rest. Do something every day to relax. Take breaks during your study time and do anything that helps you unwind.

Find Relief for Stress

Pay attention to signs of anxiety and stress. Are you nervous, worried, angry, sad? Are your muscles tense, your stomach in a knot? Is your head aching? Are you over-eating or have you lost your appetite? All these are signs of stress that can lower your success in school and interfere with your life. If you notice these signs, find a way to de-stress. Exercise and adequate rest are good for stress relief. You also might try these: stretch, take a hot bath, take a nice long shower, laugh, listen to calming music, write in a journal. If you're burdened with worries, anger, or problems, talk to someone—a good friend, or a teacher or parent or other trusted adult.

©Incentive Publications, Inc., Nashville, TN

Get a Grip (on Study Habits)

Here's some good advice for getting set to improve your study habits. Check up on yourself to see how you do with each of these tips. Then set goals where you need to improve.

. . . in school:

1. Go to class.
You can't learn anything in a class if you are not there.
Go to all your classes. Show up on time.
Take your book, your notebook, your pencil, and other supplies.

2. Choose your seat wisely.
Sit where you won't be distracted. Avoid people with whom you'll be tempted to chat.
Stay away from the back row. Sit where you can see and hear.

3. Pay attention.
Get everything you can out of each class.
Listen. Stay awake.
Your assignments will be easier
if you've really been present in the class.

4. Take notes.
Write down main points. If you hear it AND write it, you'll be likely to remember it.

5. Ask questions.
It's the teacher's job to see that you understand the material. It's your job to ask if you don't.

6. Use your time in class.
Get as much as possible of the next day's assignment finished before you leave the class.

7. Write down assignments.
Do not leave class until you understand the assignment and have it written down clearly.

8. Turn in your homework.
If you turn in every homework assignment, you are a long way toward doing well in a class—even if you struggle with tests.

Top 10 Tips for Success
(for getting better grades)

1. Get enough rest.
2. Write down your assignments.
3. Go to class (and be on time).
4. Take notes.
5. Pay attention in class.
6. Turn off the TV.
7. Do your homework.
8. Turn in your homework.
9. Don't procrastinate.
10. Ask for help.

Better Grades & Higher Test Scores / MATH
©Incentive Publications, Inc., Nashville, TN

Get Ready: Study Habits

. . . at home:

9. Gather your supplies.

Before you sit down to study, get all the stuff together that you'll need: assignment book, notebook, notes, textbook, study guides, paper, pencils, etc. Think ahead so that you have supplies for long-term projects. Bring those home from school or shop for those well in advance.

10. Avoid distractions.

Think of all the things that keep you from concentrating. Figure out ways to remove those from your life during study time. In other words, make a commitment to keep your study time uninterrupted. If you listen to music while studying, choose music that can be in the background, not the foreground of your mind.

11. Turn off the TV.

No matter how much you insist otherwise, you cannot study well with the TV on. Plan your TV time before or after study time.

12. Make phone calls later.

Plan a time for phone calls. Like TV watching, phoning does not mix with focusing on studies. The best way to avoid this distraction is to study in a room with no phone. Call your friends when your work is finished.

13. Hide the computer games.

Stay away from video games, computer games, email, and Internet surfing. Plan time for these when studies are complete, or before you settle into serious study time.

14. Know where you're going.

Review your weekly schedule and your assignment notebook. Be sure about what it is that needs to be done. Make a clear *To-Do* list for each day, so you will know what to study. Post notes on your wall, your refrigerator, or anywhere that will remind you about what things you need to get done!

15. Plan your time.

Think about the time you have to work each night. Make a timeline for yourself. Estimate how much time each task will take, and set some deadlines. This will keep your attention from wandering and keep you focused on the task.

16. Start early.

Start early in the evening. Don't wait until 10:00 P.M. to get underway on any assignment. When it's possible, start the day before or a few days before.

17. Do the hardest tasks first.

It's a good idea to do the hardest and most important tasks first. This keeps you from avoiding procrastination on the tough assignments. Also, you'll be doing the harder stuff when your mind is the most fresh. Study for tests and do hard problems early, when your brain is fresh. Do routine tasks later in the evening.

18. Break up long assignments.

Big projects, major papers, or test preparations can be overwhelming. Break each long task down into smaller ones, then take one small task at a time. This will make the long assignments far less intimidating, and you'll have more success more often. Never try to do a long assignment all in one sitting.

19. Take breaks.

Plan a break for your body and mind every 30-45 minutes. Get up, walk around, stretch, or do something active or relaxing; however, avoid getting caught up in any long phone conversations or TV shows. You'll never get back to the studying!

20. Cut out the excuses.

It's perfectly normal to want to avoid doing school work. Just about everybody has a whole list of techniques for work avoidance. And the excuses people give for putting off or ignoring it are so numerous, they could fill a whole book.

Excuses just take up your energy. In the time you waste convincing yourself or anyone else that you have a good reason for avoiding your studies, you could be getting some of the work done. If you want to be a better student, you'll need to dump your own list of excuses.

27

21. Plan ahead for long-range assignments.

Start early on long-range assignments, big projects, and test preparations. Don't wait until the night before anything is due. You never know what will happen that last day. You could be distracted, sick, or unexpectedly derailed. Get going on long tasks a long time before due date. Make a list of everything that needs to be done for a long-range assignment (including finding references and collecting supplies). Start from the due-date and work backwards. Make a timeline or schedule that sets a time to do each of the tasks on the list.

22. Don't get behind.

Keeping up is good. Many students slip into failure, stress, and hopelessness because they get behind. The best way to avoid all of these is—NOT to get behind. This means DO your assignments on time. If you DO get behind because of illness or something else unavoidable, do something about it. Don't get further and further into the pit! Talk to the teacher. Make a plan for catching up.

23. Get on top of problems.

Don't let small problems develop into big ones. If you are lost in a class, missed an assignment, don't understand something, or have done poorly on something—act quickly. Talk to the teacher, ask a parent to help, find another student who has the information. Do something to correct the problem before it becomes monumental.

24. Ask for help.

You don't have to solve every problem alone or learn everything by yourself. Don't count on someone noticing that you need help. Tell them. Use the adults and services around you to ask for help when you need it.

25. Reward yourself for accomplishments.

If you break your assignments down into manageable tasks, you'll have more successes more often. Congratulate and reward yourself for each task accomplished—by taking a break, getting some popcorn, bragging about what you've done to someone—or any other way you discover. Every accomplishment is worth celebrating!

Some awesome facts turn up in math homework!

Math Facts:

There are 336 dimples on a regulation American golf ball.

The PGA requires a golf hole to be $4\frac{1}{2}$ inches in diameter. The minimum depth is 4 inches.

By USGA rules, five minutes is the maximum amount of time allowed to search for a lost golf ball.

GET SET →

Get Familiar with Math Tools

Formulas, symbols, properties, numbers, units of measurement—these are all handy tools that will help you with math processes. If you're going to solve problems well, you'll need to have these firmly planted in your mind. The better you are with them, the easier it will be to move around in the world of mathematics!

Add the negative integer in the set intersection to the difference between pi and the square root of the decimal equivalent of $\frac{1}{2}$.
Huh?
I'd better brush up on my math tools!

Know Your Mathematical Symbols

$	dollars	$\neq$	is not equal to
¢	cents	$\approx$	is equivalent to
∅	empty set	<	less than
{ }	empty set	>	greater than
%	percent	≥	is greater than or equal to
π	pi (3.14159)	≤	is less than or equal to
°	degrees	≐	is approximately equal to
F	Fahrenheit	~	is similar to
C	centigrade	≅	is congruent to
.	point	≅	is not congruent to
√	square root	+4	positive integer
ĕ	arc	−4	negative integer
÷	divide	⟷	line
⌐	divide	——	line segment
+	add	⟶	ray
−	subtract	∠	angle
x	multiply	m∠	measure of an angle
•	multiply	△	triangle
∪	union of sets	⊥	perpendicular
∩	intersection of sets	∥	parallel
=	is equal to	a^n	a to the nth power

The 0 in 4.03 is significant.

In 0.66, the 0 is not a significant digit.

The 0's in 1,000,00 are definitely significant.

Know Your Numbers

Even Numbers — numbers that are divisible by 2

Odd Numbers — numbers that are not even

Prime Number — a number whose only factors are 1 and itself

Composite Numbers — all numbers that are not prime

Whole Number — a member of the set of numbers (0, 1, 2, 3, 4, 5...)

Fractional Number — a number that can be named in the form a/b with a and b being any numbers, with the exception that b cannot be 0

Mixed Fractional Number — a number with a whole number and a fractional number

Decimal Number — a number written with a decimal point to express a fraction whose denominator is 10 or a multiple of 10

Mixed Decimal Number — a number with a whole number and a decimal number part

Integers — the set of numbers (1, 2, 3..., -1, -2, -3...) and 0

Negative Integers — on a number line, all the numbers to the left of 0

Positive Integers — on a number line, all the numbers to the right of 0

Rational Numbers — numbers that can be written as a ratio a/b where both a and b are integers and b is not 0 (*all integers and decimals that repeat or terminate*)

Irrational Numbers — numbers that cannot be written as a quotient of two integers (*decimals that neither repeat nor terminate*)

Real Numbers — rational and irrational numbers together are the set of real numbers

Opposite Numbers — two numbers the same distance from 0 but that are on opposite sides of 0 (*3 is the opposite of –3*)

Exponential Numbers — a number with an exponent (*An exponent is a number written next to a base number to show how many times the base is to be used as a factor.*)

Digit — one number in a numeral that holds a particular place

Significant Digits — all non-zero digits and zero when it has a non-zero digit to the left of it

Get Set: Math Tools

Know Your Properties

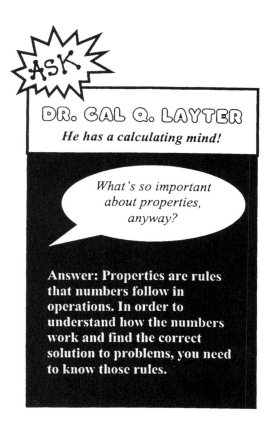

ASK

DR. CAL Q. LAYTER

He has a calculating mind!

What's so important about properties, anyway?

Answer: Properties are rules that numbers follow in operations. In order to understand how the numbers work and find the correct solution to problems, you need to know those rules.

Commutative Property For Addition – The order in which numbers are added does not affect the sum.

Example: $6 + 4 = 4 + 6$

Commutative Property For Multiplication – The order in which numbers are multiplied does not affect the product.

Example: $8 \times 3 = 3 \times 8$

Associative Property For Addition – The way in which numbers are grouped does not affect the sum.

Example: $7 + (3+2) = (7+3) + 2$

Associative Property For Multiplication – The way in which numbers are grouped does not affect the product.

Example: $(5 \times 2) \times 4 = 5 \times (2 \times 4)$

Distributive Property – To multiply a sum of numbers you may first add the numbers in parentheses and then multiply the sum.

Example: $4 \times (6+3) = 4 \times (9) = 36$

OR, you may first multiply the addends separately, then add the products.

Example: $4 \times (6+3) = (4 \times 6) + (4 \times 3) = 24 + 12 = 36$

Identity Property For Addition – The sum of 0 and any number is that number.

Example: $7 + 0 = 7$, $486 + 0 = 486$

Identity Property For Multiplication – The product of 1 and any number that is that number.

Example: $9 \times 1 = 9$, $5840 \times 1 = 5840$

Opposites Property – If the sum of two numbers is 0, then each number is the opposite of the other.

Example: -4 is the opposite of +4 because $-4 + (+4) = 0$

Zero Property – The sum of zero and any number is that number. The product of zero and any number is zero.

Example: $0 + 5 = 5$ and $5 + 0 = 5$;   Example: $0 \times 6 = 0$ and $6 \times 0 = 0$

Equation Properties – When adding or subtracting the same number or multiplying or dividing by the same number on both sides of an equation, the result is still an equation.

Examples:

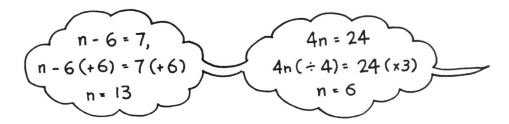

$n - 6 = 7,$
$n - 6 (+6) = 7 (+6)$
$n = 13$

$4n = 24$
$4n (\div 4) = 24 (\times 3)$
$n = 6$

Know Your Formulas

Perimeter

$P = s + s + s$	Perimeter of a triangle
$P = 2(h + w)$	Perimeter of a rectangle
$P = \text{sum of sides}$	Perimeter of irregular polygons
$C = 2\pi r$	Perimeter or circumference of a circle
$C = \pi d$	Perimeter or circumference of a circle

Coach made us run 10 laps around the perimeter of the football field. That was a 10,400-foot run!

Area

$A = \pi r^2$	Area of a circle
$A = s^2$	Area of a square
$A = bh$	Area of a parallelogram
$A = \frac{1}{2} bh$	Area of a triangle
$A = \frac{1}{2}(b_1 + b_2) h$	Area of a trapezoid

Volume or Capacity

$V = Bh$	Volume of a rectangular or triangular prism
$V = \frac{1}{3} Bh$	Volume of a pyramid
$V = s^3$	Volume of a cube
$V = \pi r^2 h$	Volume of a cylinder
$V = \frac{1}{3}\pi r^2 h$	Volume of a cone
$V = \frac{4}{3}\pi r^3$	Volume of a sphere

Can anyone tell me the volume of this football?

Get Set Tip # 1

Memorize these letters and symbols, so you will always know what they mean in formulas!

h = height
w = width
b = base
B = area of base
s = side
π = pi (3.14)
r = radius
d = diameter

Know Your Measures

The Steel Phantom Roller Coaster

People under 4 feet tall,
or 48 inches tall,
or 1 yard, 12 inches tall,
or 120 centimeters tall
or 1.2 meters tall,
or 12 decimeters tall

NOT ALLOWED!

Drats!

Length

Metric System

1 centimeter (cm)	=	10 millimeters (mm)
1 decimeter (dm)	=	10 centimeters (cm)
1 meter (m)	=	10 decimeters (dm)
1 meter (m)	=	100 centimeters (cm)
1 meter (m)	=	1000 millimeters (mm)
1 decameter (dkm)	=	10 meters (m)
1 hectometer (hm)	=	100 meters (m)
1 kilometer (km)	=	100 decameters (dkm)
1 kilometer (km)	=	1000 meters (m)

English System (U.S. Customary)

1 foot (ft)	= 12 inches (in)
1 yard (yd)	= 36 inches (in)
1 yard (yd)	= 3 feet (ft)
1 mile (mi)	= 5280 feet (ft)
1 mile (mi)	= 1760 yards (yd)

Area

Metric System

1 square meter (m²)	= 100 square decimeters (dm²)
1 square meter (m²)	= 10,000 square centimeters (cm²)
1 hectare (ha)	= 0.01 square kilometer (km²)
1 hectare (ha)	= 10,000 square meters (m²)
1 square kilometer (km²)	= 1,000,000 square meters (m²)
1 square kilometer (km²)	= 100 hectares (ha)

English System (U.S. Customary)

1 square foot (ft²)	= 144 square inches (in²)
1 square yard (yd²)	= 9 square feet (ft²)
1 square yard (yd²)	= 1296 square inches (in²)
1 acre (a)	= 4840 square yards (yd²)
1 acre (a)	= 43,560 square feet (ft²)
1 square mile (mi²)	= 640 acres (a)

Capacity

Metric System

1 teaspoon (t)	= 5 milliliters (mL)
1 tablespoon (T)	= 12.5 milliliters (mL)
1 liter (L)	= 1000 milliliters (mL)
1 liter (L)	= 1000 cubic centimeters (cm³)
1 liter (L)	= 1 cubic decimeter (dm³)
1 liter (L)	= 4 metric cups
1 kiloliter (kL)	= 1000 liters (L)

English System (U.S. Customary)

1 tablespoon (T)	= 3 teaspoons (t)
1 cup (c)	= 16 tablespoons (T)
1 cup (c)	= 8 fluid ounces (fl oz.)
1 pint (pt)	= 2 cups (c)
1 pint (pt)	= 16 fluid ounces (fl oz)
1 quart (qt)	= 4 cups (c)
1 quart (qt)	= 2 pints (pt)
1 quart (qt)	= 32 fluid ounces (fl oz)
1 gallon (gal)	= 16 cups (c)
1 gallon (gal)	= 8 pints (pt)
1 gallon (gal)	= 4 quarts (qt)
1 gallon (gal)	= 128 fluid ounces (fl oz)

Cooking With Cookie

Today's recipe is called Cookie's Surprise.

Mix 2 pt. of sour cream
1 C horseradish
4 qt. root beer
7 lbs. sweet potatoes
Boil for 3 ½ hours, and . . .

Surprise!

Volume

Metric System

1 cubic decimeter (dm³) = 0.001 cubic meter (m³)
1 cubic decimeter (dm³) = 1000 cubic centimeters (cm³)
1 cubic decimeter (dm³) = 1 liter (L)
1 cubic meter (m³)   = 1,000,000 cubic centimeters (cm³)
1 cubic meter (m³)   = 1000 cubic decimeters (dm³)

English System (U.S. Customary)

1 cubic foot (ft³) = 1728 cubic inches (in³)
1 cubic yard (yd³) = 27 cubic feet (ft³)
1 cubic yard (yd³) = 46,656 cubic inches (in³)

Better Grades & Higher Test Scores / MATH
©Incentive Publications, Inc., Nashville, TN

Get Set: Math Tools

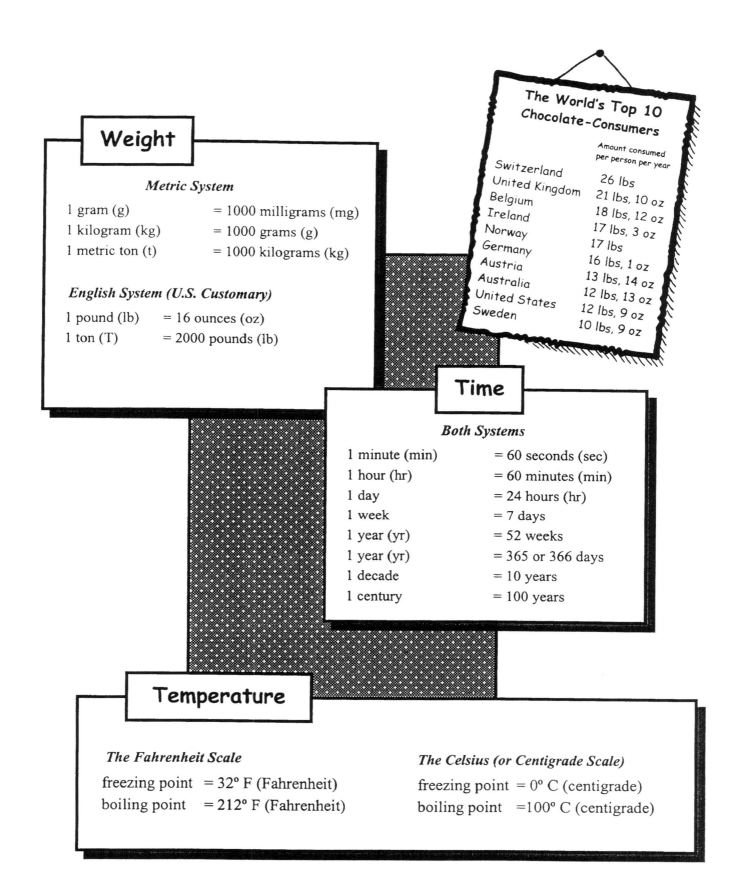

Weight

Metric System

1 gram (g)	= 1000 milligrams (mg)
1 kilogram (kg)	= 1000 grams (g)
1 metric ton (t)	= 1000 kilograms (kg)

English System (U.S. Customary)

1 pound (lb)	= 16 ounces (oz)
1 ton (T)	= 2000 pounds (lb)

The World's Top 10 Chocolate-Consumers

	Amount consumed per person per year
Switzerland	26 lbs
United Kingdom	21 lbs, 10 oz
Belgium	18 lbs, 12 oz
Ireland	17 lbs, 3 oz
Norway	17 lbs
Germany	16 lbs, 1 oz
Austria	13 lbs, 14 oz
Australia	12 lbs, 13 oz
United States	12 lbs, 9 oz
Sweden	10 lbs, 9 oz

Time

Both Systems

1 minute (min)	= 60 seconds (sec)
1 hour (hr)	= 60 minutes (min)
1 day	= 24 hours (hr)
1 week	= 7 days
1 year (yr)	= 52 weeks
1 year (yr)	= 365 or 366 days
1 decade	= 10 years
1 century	= 100 years

Temperature

The Fahrenheit Scale

freezing point	= 32° F (Fahrenheit)
boiling point	= 212° F (Fahrenheit)

The Celsius (or Centigrade Scale)

freezing point	= 0° C (centigrade)
boiling point	=100° C (centigrade)

Measurement Equivalents

From English to Metric

English Customary Unit	Approximate Metric Equivalent
inch	2.54 centimeters
foot	30.48 centimeters
yard	0.9144 meters
mile	1.609 kilometers
acre	4047 square meters
ounce	28.3495 grams
pound	453.59 grams
ton	907.18 kilograms
pint	0.4732 liters
quart	0.9463 liters
gallon	3.785 liters
bushel	35.2390 liters

From Metric to English

Metric Unit	Approximate English Equivalent
millimeter	0.04 inch
centimeter	0.39 inch
meter	39.37 inches
kilometer	3,281 feet or .62 miles
gram	0.0353 ounce
hectogram *(100 grams)*	3.53 ounces
kilogram	2.2 pounds
metric ton	2204.6 pounds or 1.1 tons
liter	1.06 quarts

The lowest temperature ever measured is –128.6° F, recorded at Vostok, Antarctica.

Temperature Conversions

To change Fahrenheit to Celsius: subtract 32, then multiply by $\frac{5}{9}$

To change Celsius to Fahrenheit: multiply by $\frac{9}{5}$, then add 32

Better Grades & Higher Test Scores / MATH
©Incentive Publications, Inc., Nashville, TN

Get Set: Math Tools

Get Sharp with Thinking Skills

Your brain is capable of an amazing variety of accomplishments! There are different levels and kinds of thinking that your brain can do—all of them necessary to get you set for good learning and studying.

To solve math problems, your brain must use many different processes. Here are some of the thinking skills that are frequently used in doing math tasks. Use this information to freshen up your mental flexibility and put these skills to use as you learn math concepts and solve problems.

Recall – To **recall** is to know and remember specific facts, names, processes, categories, ideas, generalizations, theories, or information.

This thinking skill helps you remember such things as: *the formula for finding the area of a trapezoid, the characteristics of a cylinder, the difference between a bar graph and a line graph, how to round a number, the value of a dollar, or how to bisect an angle*

Classify – To **classify** is to put things into categories. When you classify ideas, numbers, topics, or things, you must choose categories that fit the purpose and clearly define each category.

12, 6, 33, 90, 15, 27, 9, 42, 63, 84

There are many different ways to classify this group of numbers. They are all whole numbers. They are also all multiples of three, positive integers, and composite numbers.

Generalize – To **generalize** is to make a broad statement about a topic based on observations or facts. A generalization should be based on plenty of evidence (facts, observations, and examples). Just one exception can prove a generalization false.

Safe Generalization:

Positive integers have values greater than negative integers.

Invalid generalizations:

faulty generalization-A faulty generalization is invalid because there are exceptions.

Positive integers have absolute values greater than negative integers.

broad generalization- A broad generalization suggests something is *always* or *never* true about *all* or *none* of the members of a group. Most broad generalizations are untrue.

Math students always have a harder time with division than with multiplication.

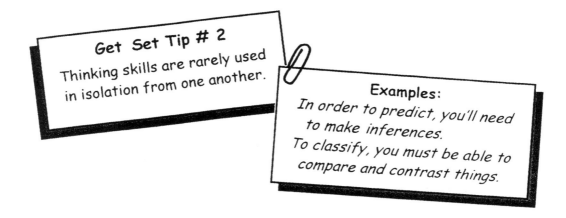

Get Set Tip # 2
Thinking skills are rarely used in isolation from one another.

Examples:
In order to predict, you'll need to make inferences.
To classify, you must be able to compare and contrast things.

Elaborate – To **elaborate** is to provide details about a situation (to explain, compare, or give examples).

When you elaborate, you might use phrases such as these: *so, because, however, but, an example of this is, on the other hand, as a result, in addition, moreover, for instance, such as, on the other hand, if you recall, furthermore, another reason is.*

Example: *Transformations are movements of geometric shapes. An example of this is a slide. In addition, flips and turns are transformations.*

Predict – To **predict** is to make a statement about what will happen. Predictions are based on some previous knowledge, experience, or understanding.

Example: *Joe knows that his mom washed all his dark socks yesterday and put them back in his drawer. He knows that most of his white socks are dirty. He predicts that when he reaches in his drawer (without looking) for a pair of socks, he'll pull out a dark pair.*

Infer – To **infer** is to make a logical guess based on information.

Example: *The snack stand at the football game made twice as much money this week as last week. The manager infers that there were more customers this week than last.*

Recognize Cause and Effect – When one event occurs as the result of another event, there is a **cause-effect relationship** between the two.

Recognizing causes and effects takes skill. When reading a math equation or problem, pay careful attention to words or symbols that give clues to cause and effect (*the reason was, because, as a result, consequently, so, therefore*).

Example: *Julia missed the sale at the ski shop, so she paid $90 more for her ski boots than Andy.*
       (cause)   (effect)

Hypothesize - To **hypothesize** is to make an educated guess about a cause or effect.

A hypothesis is based on examples that support a theory but do not prove it. A hypothesis is something that can be—and should be—tested.

If I graph the equation $x - y = -3$, the resulting figure will be a straight line.

Extend - To **extend** is to connect ideas or things together, or to relate one thing to something different, or to apply one idea or understanding to another situation.

You are extending when. . .

. . . *you see that simplifying $2x + 4x = 24$ to $6x = 24$ helps you find x, and then you use the same simplifying technique to solve a new equation, $7b - 12b + 20b = 100$.*

. . . *you learn that the chances of getting tails when you flip a penny is ½ since a penny has two sides, so you realize that the chances of getting a 6 when you toss one die is ⅙ since a die has 6 surfaces.*

Compare & Contrast

When you **compare** things, you describe similarities.
For instance, as I've shown on this graph: you're tall and I'm tall. I have one head and you have one head. I have two legs and you have two legs. I have ten fingers and you have ten fingers, and so forth.

When you **contrast** things, you describe the differences. For instance, I have charm and you don't. I'm good-looking, and you are not. I'm sensible, and you are silly, and so forth.

Draw Conclusions - A **conclusion** is a general statement that someone makes after analyzing examples and details.

A conclusion generally involves an explanation someone has developed through reasoning.

Maxie has been counting the faces on solid geometric figures. She notices that a triangular prism has fewer faces than a rectangular prism, and that a hexagonal prism has more faces than a rectangular prism.

She draws these conclusions:
1) *The shape of the base has a relationship to the number of faces.*

2) *As the number of sides on the base figure increases, the number of faces on the solid figure will increase.*

Analyze - To **analyze,** you must break something down into parts and determine how the parts are related to each other and how they are related to the whole.

For instance, you must analyze to . . .

   . . . *evaluate a mathematical expression*

   . . . *understand a word problem.*

   . . . *write a numeral in expanded notation.*

   . . . *find a common denominator for two unlike fractions.*

Synthesize - To **synthesize,** you must combine ideas or elements to create a whole.

For instance, you must synthesize to . . .

   . . . *solve an equation.*

   . . . *combine facts you've learned to solve a long division problem.*

   . . . *simplify a mathematical expression.*

   . . . *create a graph from statistical data.*

Think Logically (or Reason) - When you think **logically,** you take a statement or situation apart. You use **inductive or deductive** reasoning to examine the details that support a conclusion, or the generalization, that leads to specific details.

Four hikers are plodding along a trail.
The whistling hiker wears long pants.
One hiker has blisters. She wears no hat.
Only one hiker has mosquito bites.
A hiker with mosquito bites wears a hat.
A hiker with a swollen ankle is one of two hikers
   wearing long pants.
A hiker with bruised shoulders is wearing a hat.
The hiker with blisters has no other bothersome ailments.
The hiker with mosquito bites has no other bothersome
   ailments.
Only two hikers are wearing hats.
Only one hiker is whistling.

After reading the above information, I logically reason the following:
   Three hikers have no mosquito bites.
   The hiker with a swollen ankle is not wearing a hat.
   The hiker with mosquito bites might be wearing shorts.
   The hiker with blisters is not the one with the swollen ankle.
   The whistling hiker could have a swollen ankle.

Get Polished in Problem-Solving Skills

These skills are absolute musts for success at solving math problems. Don't go near math problems without them! Practice them until you have them finely polished.

1. Read the directions twice.

It's a good idea to do this twice, since your eyes sometimes play tricks on you.

2. Read the problem carefully.

Don't ever assume that you know what the assignment is. Don't rush ahead to do a math problem until you read the instructions.

This is tricky. The question is not what I thought it would be!

3. Identify the problem to be solved.

Find out EXACTLY what you need to find out! Sometimes there are several problems that could be solved with a set of information. But you need to find what you are asked to find! (Underline the question.)

4. Figure out what information is important.

Some facts or information in a problem may not be necessary to solve the question that is asked. Learn to separate the necessary information from the unnecessary. (Actually "X" out the facts that are not needed.)

I'll cross out this fact. It isn't even needed.

5. Decide on the operation needed.

Will you add, subtract, multiply, or divide? Will you need to use more than one operation? Look for words that give clues as to what operation(s) might be needed.

This answer doesn't make sense! The area of half the pizza can't be bigger than the area of the whole pizza!

6. Choose a strategy.

Different strategies are useful for different problems. Choose a strategy that makes sense for the kind of problem you have to solve.

7. Draw, scratch, doodle, diagram.

Draw diagrams, write equations, make charts, draw pictures. Do whatever writing you need in order to figure out the answer.

8. Look closely at the answer.

Does it make sense? Is it a logical answer to the question? Use another operation or strategy to check the accuracy of your solution.

Clues to Operations

In word problems, watch carefully for these words and phrases. They will give you clues to the operation needed to solve the problem.

These words suggest addition:

> *total*
> *sum*
> *add up to*
> *in all*
> *together*
> *both*
> *increased by*
> *all together*

These words suggest subtraction:

> *difference*
> *less than*
> *decreased by*
> *left over*
> *remain*
> *take away*
> *have left*
> *fewer than*
> *how much more*
> *how much less*
> *change (in money problems)*

These words suggest multiplication:

> *times*
> *how many times*
> *a product of*
> *multiplied by*
> *twice as much as*

These words suggest division:

> *divided by*
> *what was the average*
> *half as much*
> *any fraction*
> *parts*
> *equal parts*
> *sharing*
> *split up or cut up*
> *equally distributed among*

What is the exact question to be answered?

What facts are needed?

What facts are not needed?

What operations will I need to use?

What is a good strategy for solving this problem?

Does my answer make sense??

How can I check my solution to see if it is accurate?

Ask questions like these when you are trying to solve a problem.

Better Grades & Higher Test Scores / MATH
©Incentive Publications, Inc., Nashville, TN

Get Set: Probem-Solving Skills

Spectators drove a total of 937,500 miles to watch the skating competition. ~~They came from 68 different cities. Some traveled as long as 40 hours.~~ There were 12,500 spectators at the competition. (On the average) how far did each spectator travel?

$$\begin{array}{r} 75 \\ 12,500 \overline{)937,500} \\ -87500 \\ \hline 62500 \\ -62500 \\ \hline 0 \end{array}$$

$$\begin{array}{r} 12,500 \\ \times\ 75 \\ \hline 62500 \\ +87500 \\ \hline 937,500 \end{array}$$

75 mi average

Raja's sister Tarai is 4 years older than Raja. In 14 years, the sum of their ages will be 40. (How old is Tarai now?)

$$R + 14 + \overset{Tarai}{(R+4)} + 14 = 40$$

$$2R + 32 = 40$$
$$2R = 40 - 32$$
$$2R = 8$$
$$R = 4 \ (Raja)$$
$$T = 4 + 4 \ (Tarai)$$

Tarai is now 8 years old.

Notice that Suki underlined the question she would need to answer.

She also crossed out the facts that were not needed to solve the problem.

Suki circled the words "on the average." These words gave her a clue that she would need to divide.

When she finished the division problem, she checked her answer with a multiplication problem.

Andy circled the question that he needs to answer.

He took the facts given and turned them into an equation.

When he solved the equation, he found Raja's age. But the problem asked him to find Tarai's age. So he had another step to do.

Better Grades & Higher Test Scores / MATH
©Incentive Publications, Inc., Nashville, TN

Notice that Charlie circled the part of the directions that told him exactly what to find.

He realized he would not be able to write a number problem to solve this problem. He would need to use logic.

He drew a diagram to help him organize the facts and find the answer.

Four neighbors in Sweetholm live in a row of houses numbered 20, 21, 22, and 23.

Dr. Caramel's house number is higher than Felix's number.

Felix Fondant does not live next to Judge Fudge.

Officer Brittle lives on one end of the row.

Dr. Caramel lives between Felix and the judge.

The resident in # 22 has a pet scorpion.

Judge Fudge's house number is lower than Officer Butler's.

(Who owns the scorpion?)

| 20 | 21 | 22 | 23 |
| Felix Fondant | Dr. Caramel | Judge Fudge (scorpion) | Off. Brittle |

Judge Fudge owns the scorpion.

Lori circled the question to be answered.

Then she chose a strategy for solving the problem—trial and error.

She wrote down each of the 3-digit multiples of 9 between 400 and 500.

Then she examined each of them and eliminated each number that did not fit all the descriptions.

An even, 3-digit number is a multiple of 9.
It is > 400 and < 500.
All the digits in the number are different.
No digit is > 4.
The digit in the tens place is an odd number.
(What is the number?)

~~405~~   ~~491~~   ~~477~~
~~414~~   ~~450~~   486
~~423~~   ~~459~~   ~~495~~
(432)   ~~468~~

The number is 432

Get Serious about Study Skills

Better Listening

Keep your ears wide open! You can increase your understanding of math concepts and processes if you listen well. Here are some tips for smart listening. They can help you get involved in with the information instead of letting it just buzz by your ears.

Get Set Tip # 3

Stop talking! (You can't listen while you talk.)

1. Realize that the information is important.

Here's what you can get when you listen to someone who is talking to you about math:

. . . details about how a math process works.

. . . help solving problems.

. . . examples of problems solved correctly.

. . . hazards or difficulties you might face when doing a particular process.

. . . hints for how to solve certain kinds of problems or succeed with certain processes.

. . . meanings of terms used in problems.

. . . directions for certain assignments.

Did she really say what I think she said?

2. Be aware of the obstacles to good listening.

Know ahead of time that these will interfere with your ability to listen well. Try to avoid them, alter them, or manage them so they don't get in the way.

. . . tiredness

. . . surrounding noise

. . . uncomfortable setting

. . . personal concerns, thoughts, or worries

. . . wandering attention

. . . too many things to hear at once

. . . missing the beginning or ending

. . . talking

3. Make a Commitment to Improve

You can't always control all obstacles (such as the comfort of the setting or the quality of the speaker's presentation), but there are things you can control. Put these to work to gain more from your listening.

. . . Get enough rest.

. . . Do your best to be comfortable while you listen.

. . . Cut out as many distractions as possible. Keep your mind focused on what is being said.

. . . Look directly at the speaker.

. . . Take notes. Write down sample problems the speaker shows or solves.

. . . As the speaker talks, think of examples or relate the information to your life.

. . . Don't miss the beginning or ending. Pay special attention to opening and closing remarks.

. . . Pay special attention to anything that is repeated.

. . . As soon as possible after listening, summarize or review what you have heard.

46

Better Grades & Higher Test Scores / MATH
©Incentive Publications, Inc., Nashville, TN

Careful Reading

There is plenty of reading in math. Textbooks and other learning materials explain math processes. Many math problems are more than just numbers—they include written information you need to understand and interpret. Almost all problems include some sort of instructions to follow. So, to succeed in math, you need to make good use of reading skills.

Before you read a math assignment of problem, have a clear idea of the purpose for reading. Are you reading to find directions for the assignment? Are you reading to learn how to do a math operation or process? Are you reading to solve word problems? In all these cases above, you need to read closely and carefully. To learn or review a process, or to solve a problem, you will need to read the information through more than once.

Draw a diagram to solve the problem below. Write a clear answer to the question. Show your diagram and explain how you arrived at your answer.

The boat race began after lunch on a sunny Friday. At this point in the race, four boats are out in front of the rest. The Cincinnati Speed Kings' boat is in the lead. The Miami Splash team is not next to the Atlanta Racers. The Washington Wizards are not in second place. The Seattle Stingrays' boat is in fifteenth place. One hour into the race, it started raining. Could the Washington Wizards currently be in fourth place?

If you read the problem (above) carefully, you should be able to . . .

. . . find clear directions for the assignment.

> *Draw a diagram to solve the problem.*
> *Write a clear answer to the question.*
> *Show your diagram and explain how you arrived at your answer.*

. . . identify the problem or question that needs to be solved.

> *Is it possible that the Washington Wizards are currently in fourth place?*

. . . find the information need to solve the problem.

> *At this point in the race, four boats are out in front of the rest.*
> *The Speed Kings' boat is in the lead.*
> *The Miami Splash Team is not next to the Atlanta Racers.*
> *The Washington Wizards are not in second place.*

. . . identify and skip over any information not necessary for solving the problem.

> *The boat race began after lunch on a sunny Friday.*
> *One hour into the race, it started raining.*
> *The Seattle Stingrays' boat is in fifteenth place.*

Taking Notes

Get Set Tip # 4

Take good notes! You'll remember the material better with less studying.

A study skill of major importance is knowing how to take notes well and use them effectively. Good notes from classes and from reading are valuable resources to anyone who's trying to do well as a student. A lot of learning goes on while you're taking notes—you may not even realize it's happening!

Here's the basic process for taking notes:

> I'd better write this down.

> I've got a sample of each kind of problem.

> Well, my notes are the best.

| 1. Listen or read for a main idea, key formula, or important process. Write it down. | 2. Write examples, problems, definitions, and details to support each idea key. | 3. Review your notes as soon as possible after the class. This will help to fix the information in your brain. |

Here's what is happening when you take notes:

1. When you take notes, you naturally listen better. (You have to listen to get the information to write it down!)

2. When you listen for the purpose of taking notes, you naturally learn and understand the material better. Taking notes forces you to focus on what's being said or read.

3. When you sort through the information and decide what to write, you naturally think about the material and process it—making it more likely that you'll remember it.

4. The actual act of writing the notes fixes the information more firmly in your brain.

5. Having good notes in your notebook makes it possible for you to review and remember the material. Having written examples of the problems, formulas, or math processes makes it possible for you to see the math done right. This makes reviewing and remembering it easier.

Tips for Wise Note-Taking

in class . . .

- Have a notebook or a notebook section for math.
- When the class begins, write the topic for the day at the top of a clean page.
- Write the date at the top of the page.
- Only take notes on one side of the paper.
- Use an erasable pen for clear notes, not a pencil.
- Write down examples of problems and problem solutions.
- Write notes to yourself about how the problem was solved.
- Write neatly—so you can read it later.
- Leave sizable margins to the left of the outline.
- Use these spaces to star important items or write key words.
- Leave a blank space after each main idea section.
- Pay close attention to the opening and closing remarks.
- Listen more than you write.
- ASK about anything you do not understand.

Get Set Tip # 5
When you take notes in class, be alert for signals from the teacher about important ideas.
Write down anything the speaker (or teacher)...
...writes on the board.
...gives as a definition.
...emphasizes with his or her voice.
...repeats.
...says is important.

. . . from a textbook assignment

- Skim through one section at a time to get the general idea. (Use the textbook divisions as a guide to separate sections, or read a few paragraphs at a time.)
- Then go back and write down the main ideas.
- For each main idea write a few supporting details or examples. If a math operation or process is explained, write down a sample problem solved correctly.
- Notice bold or emphasized words or phrases. Write these down with definitions.
- Read captions under pictures. Pay attention to facts, tables, charts, graphs, and pictures, and the explanations that go along with them. Put information in your notes if it is very important.
- Don't write too little. You won't have all the main points or enough examples.
- Don't write too much. You won't have time or interest in reviewing the notes.

How to Prepare for a Test

Good test preparation does not begin the night before the test.

The time to get ready for a test starts long before this night.

Here are some tips to help you get ready – weeks before the test and right up to test time.

1. **Start your test preparation at the beginning of the year— or at least as soon as the material is first taught in the class.**

    The purpose of a test is to give a picture of what you are learning in the class. That learning doesn't start 12 hours before the test. It starts when you start attending the class. Think of test preparation this way, and you'll be able to be less overwhelmed or anxious about an upcoming test.

    You'll be much better prepared for a test *(even one that is several days or weeks away)* if you . . .

    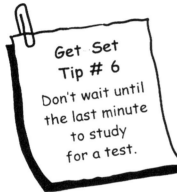

    . . . pay attention in class.

    . . . take good notes and work out sample problems.

    . . . keep your notes and class handouts organized.

    . . . read all your assignments.

    . . . do your homework regularly.

    . . . make up any work you miss when you're absent.

    . . . ask questions in class about anything you don't understand.

    . . . review notes and handouts regularly.

2. **Once you know the date of the test, make a study plan.**

    Look over your schedule and plan time to start organizing and reviewing material.

    Allow plenty of time to go through all the material.

    Your brain will retain more if you review it a few times and spread the studying out over several days.

3. **Get all the information you can about the test.**

    Write down everything the teacher says about the test.

    Get clear about what material will be covered.

    If you can, find out about the format of the test.

    Make sure you get all study guides the teacher distributes.

    Make sure you listen well to any in-class reviews.

4.   Use your study time effectively.

Don't:	Do:
• spend your study time blankly staring at your notebook or mindlessly leafing through your textbook.	• gather and organize all the notes and handouts you have.
• study with someone else unless that person actually helps you learn material better.	• review your text, pay attention to bold words, bold statements, and examples of concepts, operations, or problems.
• study in blocks of time so long that you get tired, bored, or distracted.	• identify the kinds of problems in the section being tested; practice solving a few of each kind.

Do:

• gather and organize all the notes and handouts you have.

• review your text, pay attention to bold words, bold statements, and examples of concepts, operations, or problems.

• identify the kinds of problems in the section being tested; practice solving a few of each kind.

• review the questions at the end of text sections; practice answering them.

• review your notes, using a highlighter to emphasize important points.

• review the study guides provided by the teacher.

• review any previous quizzes on the same material.

• predict the questions that may be asked and kinds of problems that will be included; think about how you would answer them.

• make study guides and aids for yourself.

• make sets of cards with key vocabulary words, terms and definitions, main concepts, and types of problems.

• ask someone (reliable) to quiz you on the main points and terms.

Now that I've got all my supplies organized, I'm finally ready to go to bed.

5.   Get yourself and your supplies ready.

Do these things the night before the test (not too late):

> *Gather all the supplies you need for taking the test (good pencils with erasers, erasable pens, scratch paper, calculator with batteries).*
>
> *Put these supplies in your school bag.*
>
> *Gather your study guides, notes, and text into your bag.*
>
> *Get a good night of rest.*

In the morning:

> *Eat a healthy breakfast.*
>
> *Look over your study guides and note card reminders.*
>
> *Relax and be confident that your preparation will pay off.*

How to Take a Test

Before the test begins

- Have supplies ready: take sharpened pencils, scratch paper, calculator, eraser.
- Try to get a little exercise before class to help you relax.
- Go to the bathroom and get a drink.
- Arrive at the class on time (or a bit early).
- Get settled into your seat; get your supplies out.
- If there's time, you might glance over your study guides while you wait.
- To relax, take some deep breaths; exhale slowly.

When you get the test

- Put your name on all pages.
- Before you write anything, scan the test to see how long it is, what kinds of questions it has, and generally what it includes.
- Think about your time and quickly plan how much time you can spend on each section.
- Read each set of directions twice. Circle key words in the directions.
- Answer all the short-answer questions. Do not leave any blanks.
- If you are not sure of an answer, make a smart guess.
- Don't change an answer unless you are absolutely sure it is wrong.

Get Set Tip # 7

Research shows that your first answer is correct more often than not! So stick with it unless you are sure you know the right answer.

Maxie is not sure of the answer so she makes a smart guess. She puts an **X** by the problem so she will remember to come back to it later.

I think it's **C**.

X

14. 32 (7 – 4) = (3 x 7) – (3 x 4)

This is an example of
a. the associative property.
b. the identity property.
c. the distributive property.
d. the commutative property.

15. a + 10 b = 5.

If b = 10, then a =
a. -95
b. 125

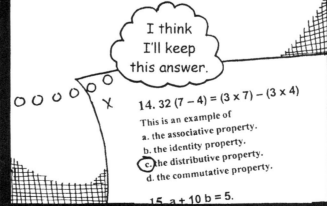

When she comes back to the question, she is still not sure, so she stays with her first answer.

I think I'll keep this answer.

X

14. 32 (7 – 4) = (3 x 7) – (3 x 4)

This is an example of
a. the associative property.
b. the identity property.
c. the distributive property.
d. the commutative property.

15. a + 10 b = 5.

Better Grades & Higher Test Scores / MATH
©Incentive Publications, Inc., Nashville, TN

More Test-Taking Tips

Tips for Solving Word Problems

A **word problem** uses words to describe a problem or question which needs a solution.

- First, read through the problem twice.
- Identify the question to be answered. Underline it.
- Circle key facts needed to solve the problem.
- Circle clue words that point to the correct operation.
- Choose a strategy for solving the problem.
- Write down a problem or equation that could solve the problem.
- Draw diagrams, charts, or pictures if you need them.
- Solve the problem.
- Go back and read the problem again. Ask yourself:

  Did I answer the question that the problem required?
  Is this answer reasonable?

- Check your answer using another method or strategy, if you have time.

Get Set
Tip # 8
On all questions,
on all tests —
always read
the directions
through twice.

Tips for Solving Number Sentences or Equations

A **number sentence or equation** is a problem made up of numbers, usually with a missing number for you to find. In an equation, the missing number is often represented by a letter.

- Read the number sentence twice.
- Identify the missing element or number that you need to find.
- Simplify the sentence or equation by combining elements that are the same, or by doing easy computations.
- Identify the operation needed to solve the problem.
- Solve the problem.
- Take your answer and write it into the number sentence or equation.
- Read through the problem again to make sure it is correct with your answer inserted.

Better Grades & Higher Test Scores / MATH
©Incentive Publications, Inc., Nashville, TN

Get Set: Study Skills

Even More Test-Taking Tips

Tips for Answering Multiple Choice Questions

Multiple choice questions give you several answers from which to choose.

- Read the question through twice.
- Before you look at the choices, close your eyes and answer the question. Then look for that answer.
- Read all the choices through before you circle one.
- If you are not absolutely sure, cross out answers that are obviously incorrect.
- Choose the answer that is most complete or most accurate.
- If you're not absolutely sure, choose an answer that has not been ruled out.
- Do not change an answer unless you are absolutely sure of the correct answer.

Tips for Answering Matching Questions

Matching questions ask you to recognize facts or definitions in one column that match facts, definitions, answers, or descriptions in a second column.

- Read through both columns to familiarize yourself with the choices.
- Do the easy matches first.
- Cross off answers as you use them.
- Match the left-over items last.
- If you don't know the answer, make a smart guess.

Tips for Answering Fill-in-the Blank Questions

Fill-in-the-blank questions ask you to write a word that completes the sentence.

- Read through each question. Answer it the best you can.
- If you don't know an answer, **X** the question and go on to do the ones you know.
- Go back to the **X**'d questions. If you don't know the exact answer, write a similar word or definition—come as close as you can.
- If you have no idea of the answer, make a smart guess.

Tips for Answering True-False Questions

True-False questions ask you to tell whether a statement is true or false.

- Watch for words like *most, some,* and *often.* Usually statements with these words are TRUE.
- Watch for words like *all, always, only, never, none, nobody,* and *never*. Usually statements with these words are FALSE.
- If any part of a statement is false, then the item is FALSE.

GET SHARP →

on

Number Concepts & Relationships

My favorite set!

Numbers & Sets

A number is an idea!
It is a mathematical idea about the amount contained in a set.

A numeral is a symbol that represents a number.

A set is a collection of items called members or elements. The set below has 7 members or elements.

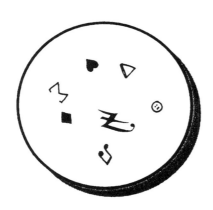

A Venn diagram can be used to show the members of a set. (See diagram above.)

Here is another way to show the members of a set:

$$\{ \blacklozenge , \heartsuit , \sharp , \odot , \maltese , \triangle , 3 \}$$

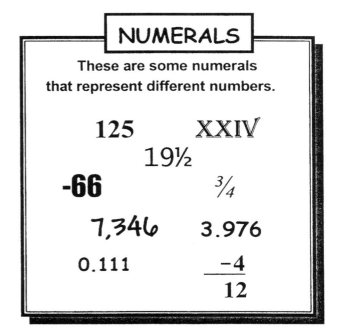

NUMERALS

These are some numerals that represent different numbers.

125   XXIV
            19½
-66   ¾

7,346   3.976

0.111   −4
            ‾‾‾
              12

A finite set has a specific number of members. The set shown above is a finite set.

An infinite set has an unlimited number of members.

An empty set has no elements. It is also called a null set.

The symbols $\{ \ \}$ and $\varnothing$ represent an empty set.

Subsets are sets made of any member of a set or any combination of members of a set.

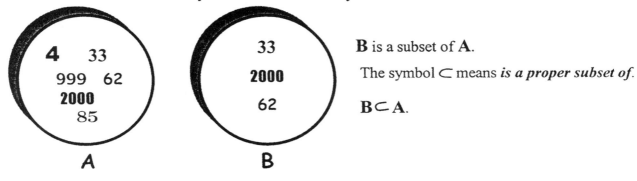

B is a subset of A.

The symbol $\subset$ means *is a proper subset of*.

$B \subset A$.

Equivalent sets are sets having the same number of members.

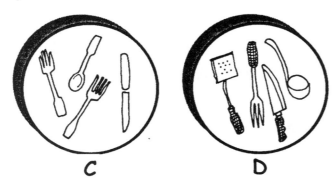

C and D are equivalent sets.

Intersection of sets is the set of members common to each of two or more sets..

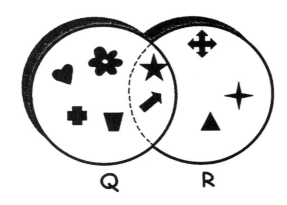

The symbol $\cap$ represents the intersection of sets.

$Q \cap R$.

The intersection of sets Q and R is

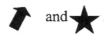 and ★ .

Union of sets is a set containing the combined members of two or more sets.

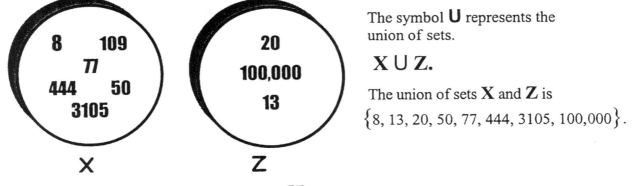

The symbol $\cup$ represents the union of sets.

$X \cup Z$.

The union of sets X and Z is

$\{8, 13, 20, 50, 77, 444, 3105, 100,000\}$.

The Roman Numeral System

Number systems have been around for a very long time back in the days of the cave dwellers, people counted belongings and kept track of the calendar. One ancient system dates back over 2000 years, to the Roman Empire.

In the Roman system, seven letters represent numbers.

I	=	1
V	=	5
X	=	10
L	=	50
C	=	100
D	=	500
M	=	1000

Other numerals are made from combinations of the letters.

$4 = 5 - 1$ *(one less than 5)* or **IV**

$9 = 10 - 1$ **IX**

$40 = 50 - 10$ **XL**

$90 = 100 - 10$ **XC**

$400 = 500 - 100$ **CD**

$900 = 1000 - 100$ **CM**

Roman numerals aren't really used any more, are they?

Sure! You can find them on some clocks, in outlines, on pages in books, and engraved in some buildings.

ROOM VI

Can you read these numbers?

XIX	(19)	**MDCXCVII**	(1697)
XXXIV	(34)	**MCMXCIX**	(1999)
CDLXXXIV	(484)	**MMII**	(2002)
CMXLII	(942)	**MMCMXC**	(2990)

58

The Decimal System

The word *decimal* comes from a Latin root *decem* that means *ten.* A decimal system of numbers uses ten symbols. Today these symbols are the numerals 1-9 and 0. The decimal system we use today is the Arabic system. The ten symbols used are the Arabic numerals 1, 2, 3, 4, 5, 6, 7, 8, and 9.

The decimal system contains different sets of numbers *(counting numbers, whole numbers, odd and even numbers, prime and composite numbers, integers, and rational numbers).*

THE REAL TOP 10

Counting Numbers

The set of **counting numbers** begins with 1 and continues into infinity. Sometimes these are called *natural numbers*. They are used for counting.

You can count by ones:
1, 2, 3, 4, . . .   OR   867, 868, 869, 870, . . .

Or you can skip count:
(skipping over numbers in any regular interval you choose)

by fives
5, 10, 15, 20, 25, 30. . .   OR   320, 325, 330, 335, 340, . . .

by tens
10, 20, 30, 40, 50, . . .   OR   670, 680, 690, 700, 710, . . .

by fifties
50, 100, 150, 200, . . .   OR   6250, 6300, 6350, 6400, . . .

by hundreds
1300, 1400, 1500, . . .   OR   3800, 3900, 4000, 4100, . . .

> You can count backwards, too:
> 1000, 999, 998, 997, 996, 995, . . .
> 88, 86, 84, 82, 80, 78, 76, 74, . . .
> 500, 475, 450, 425, 400, 375, . . .
> 730, 720, 710, 700, 690, 680, 670, . . .

FACT FANTASTICO

When you count by nines, the sum of all the digits in each sum = nine!
9 + 9 = 18
(1 + 8 = 9)
9 + 9 + 9 = 27
(2 + 7 = 9)
9 + 9 + 9 + 9 = 36
(3 + 6 = 9)
9 + 9 + 9 + 9 + 9 = 45
(4 + 5 = 9)
and so on!

Whole Numbers

Whole numbers make up a set of numbers within the decimal system. The set of **whole numbers** includes 1 and all the numbers that follow it into infinity, PLUS the number zero (0).

99,899...99,900...99,991...99,992...99,993...99,994...99,995...99,996...99,99

Even & Odd Numbers

Even–Steven and Odd Maude explain . . .

Even numbers are the number zero (0) and all the numbers that can be evenly divided by 2 (including the number 2 itself).

Odd numbers are all the numbers that cannot be evenly divided by 2.

That's odd!

Listen Maude,
The infield of a baseball diamond is an **EVEN 90** feet by **90** feet.
The pitcher may pitch **4** balls before walking the hitter.
Kerry Woods (of the Chicago Cubs) had a GS of **120** during his famous **20**-strikeout game—the highest GS recorded to that date. Barry Bonds (of the San Francisco Giants) earned an EVEN **$18,000,000** during the **2002** season.

There are **9** players on a baseball team and **9** innings in a game. If a hitter gets **3** strikes, he's out!
In **2001**, there were **199** new players in the major leagues. **55** players were born outside the U.S.A.
In the World Series, there have been **3** tie games, **105** shut-outs, **1** perfect game and **11** situations where a team lost the first two games and rallied to win the championship. . .

. . . and, Steven, this book set us back **$67**.

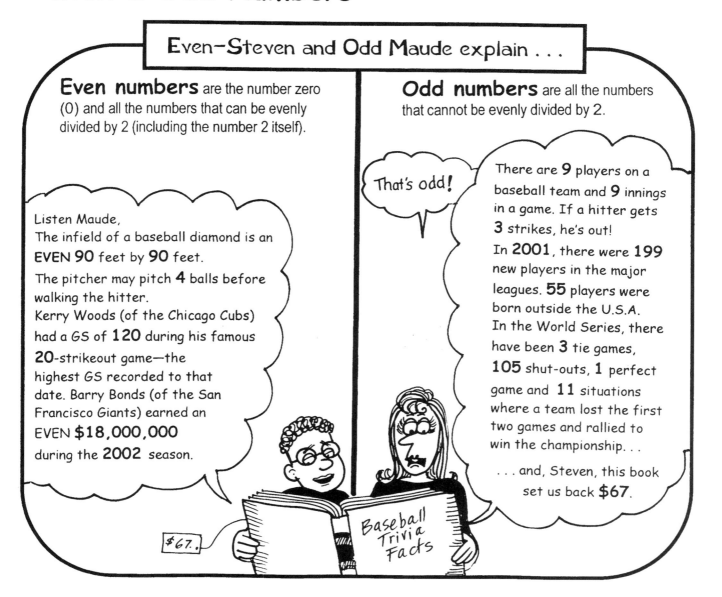

$67.

Baseball Trivia Facts

Prime & Composite Numbers

The set of prime numbers includes any whole number that has exactly two factors,—1 and itself.

Examples: 7 has only two factors (1 and 7)

83 has only two factors (1 and 83)

The set of composite numbers includes any whole number that has more than two factors.

Examples: 12 has 6 factors (1, 2, 3, 4, 6, 12)

56 has 8 factors (1, 2, 4, 7, 8, 14, 28, 56)

Get Sharp
Tip # 1

The number 1 is neither prime nor composite. It is not prime because it does not have two factors. It is not composite because it is not divisible by other numbers.

This chocolate cake is prime!

It's a fact! There are exactly 25 **prime numbers** between 1 and 100. Here they are: 2, 3, 5, 7, 9, 11, 13, 17, 19, 23, 29, 31, 37, 41, 47, 53, 59, 61, 67, 71, 73, 79, 83, 89, and 97!

Food time is prime time.

There are 73 numbers between 1 and 100 that are not prime. These are called **composite numbers**.

Can you get **prime** rib from a cow that is four years old?

Place Value for Whole Numbers

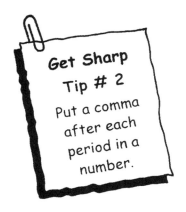
Every number has one or more **_digits_** (individual symbols).
The value of each digit depends on its place in the number.

Each digit in a number has a value 10 times greater than the digit to its right.

999

The nine on the right end is worth 9 ones.
The next nine to the left is worth 10 x 9 or 90.
The next nine to the left is worth 10 x 90 or 900.

Digits in numbers are grouped in **periods.**
The periods will help you remember the value of the different places.

Billions Period			Millions Period			Thousands Period			Ones Period		
hundred billions	ten billions	one billions	hundred millions	ten millions	one millions	hundred thousands	ten thousands	one thousands	hundreds	tens	ones
100,000,000,000	10,000,000,000	1,000,000,000	100,000,000	10,000,000	1,000,000	100,000	10,000	1,000	100	10	1

The value of the digit 4 is four hundred thousand.　　　The value of the 5 is five hundred million.

3,506,487,210

The value of the 7 is seven thousand.　　　There is an 8 in the ten thousands place.

The ten millions place and the ones place have no value.
Zero holds these places.

A scientist collected 36,789 scorpions.

Each of the scorpions had 12 eyes.

That's 441,468 eyes!

In that number, the hundred thousands place, the ten thousands place, and the hundreds place all have a value of 4!

Reading and Writing Whole Numbers

Read numbers from left to right. At each comma, say the name of the period.

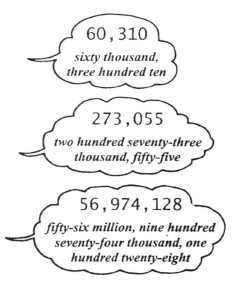

60,310
*sixty thousand,
three hundred ten*

273,055
*two hundred seventy-three
thousand, fifty-five*

56,974,128
*fifty-six million, nine hundred
seventy-four thousand, one
hundred twenty-eight*

Standard Notation

Standard notation is the ordinary numeral form of a number.

In standard notation, the number:

*five hundred ninety-six thousand,
four hundred twenty-three*

is written

596, 423.

Expanded Notation

There is another way of writing a number, one that shows the value of each digit in an expanded form. This is called **expanded notation**.

$$596,423 = 500,000 + 90,000 + 6,000 + 400 + 20 + 3.$$

OR

$$596,423 = (5 \times 10^5) + (9 \times 10^4) + (6 \times 10^3) + (4 \times 10^2) + (2 \times 10^1) + (3 \times 10^0)$$

In April of 2002, the U.S. Census Bureau reported that the world population was 6,219,866,485.

Let's see, that's. . .

six billion,
two hundred nineteen million,
eight hundred sixty-six thousand,
four hundred eighty-five.

or,

6,000,000,000 + 200,000,000 + 10,000,000 + 9,000,000 + 800,000 + 60,000 + 6,000 + 400 + 80 + 5!

Rounding Whole Numbers

Rounding is increasing or decreasing a number to the nearest ten, hundred, thousand, or any other multiple of ten. Rounding makes it easier to work with numbers.

1,533 rounded to the nearest ten is 1,530.

1,533 rounded to the nearest hundred is 1,600.

1,533 rounded to the nearest thousand is 2,000.

A 10th-grader, trying to set a world record, went for 15,843 minutes without sleep.

Rounded to the nearest ten, that's 15,840.
Rounded to the nearest hundred, that's 15,800.
Rounded to the nearest thousand, that's 16,000.
Rounded to the nearest ten thousand, that's 20,000.

Some kids decided to make a paperclip chain as long as the Great Wall of China. They figure they will need about 88,704,000 paper clips.

Rounded to the nearest ten thousand, that's 88,700,000.
Rounded to the nearest hundred thousand, that's 88,700,000.
Rounded to the nearest million, that's 89,000,000.
Rounded to the nearest ten million, that's 90,000,000.

Better Grades & Higher Test Scores / MATH
©Incentive Publications, Inc., Nashville, TN

Estimating

To **estimate** means to make a reasonable guess. Estimation is a good tool to use when you do not need to end up with a precise count or answer. Rounding is helpful when you estimate.

Suki is the manager for the basketball team. One of her jobs is making sure there are enough clean towels for showers after practices and games. She doesn't have to wash the towels, but each week she counts and folds enough clean towels and brings them to the locker room. The team has 18 players. They practice 5 times a week and play 2 games a week for 12 weeks. About how many towels will Suki fold over the season?

Round 18 players to 20.
Round the 7 practices and games to 10.
Round the 12 weeks to 10.
Multiply 20 x 10 x 10.
The estimated number of towels is 2000.

For lunch, each team member ate two hamburgers, an order of fries, a large drink, and a sundae. The 18 team members had a total of $275.00 between them. Is that enough to pay for their lunch?

$3.65 rounds to $4.00. Two hamburgers cost about $8.00. The fries were close to $2.00; each drink was about $1.00; and each sundae was about $2.00. This is a total of $13 per person. Round the team to 20 players. 20 x 13 is $260.00. $275 should be enough.

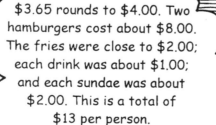

MENU

Hamburgers	$ 3.65
Hot Dogs	$ 2.50
Chicken Sandwich	$ 4.15
Fries	$ 1.75
Ice Cream Cone	$ 2.50
Ice Cream Sundaes	$ 2.20
Drinks:	
Small	$ .75
Medium	$ .95
Large	$ 1.25
Shakes	$ 3.25

Get Sharp: Kinds of Numbers

Factors

Factors are two numbers that can be multiplied together to find a product. Every whole number greater than 1 has at least 2 factors (2 numbers that, when multiplied together, yield that number.)

3 and 7 are factors of 21, because 3 x 7 = 21.

The factors of 24 are 1, 2, 3, 4, 6, 8, 12, and 24.

Common Factors

Common factors are factors that two (or more) different numbers have in common.

Get Sharp
Tip # 4
One (1)
is a factor of
every number.

The factors of 30 are   The factors of 54 are
1, 2, 3, 5, 6, 10, 15, 30   1, 2, 3, 6, 9, 18, 27, 54

The **common factors** of 27 and 54 are 1, 2, 3, 6.

Greatest Common Factors (GCF)

The greatest common factor (GCF) of two or more numbers is the common factor with the greatest value.

The factors of **20** are 1, 2, 4, 5, 10, and 20.

The factors of **28** are 1, 2, 4, 7, 14, and 28.

The factors of **36** are 1, 2, 3, 4, 6, 9, 12, 18, and 36.

The **common factors** of **20, 28,** and **36** are **1, 2, 4.**

The **greatest common factor** (GCF) is **4.**

Better Grades & Higher Test Scores / MATH
©Incentive Publications, Inc., Nashville, TN

Multiples

A **multiple** of a number is any product of that number and a whole number. Each number has an infinite number of multiples.

Some of the multiples of 7 are:

$$0 \times 7 = 0$$
$$2 \times 7 = 14$$
$$3 \times 7 = 21$$
$$4 \times 7 = 28$$
$$5 \times 7 = 35$$
$$10 \times 7 = 70$$
$$30 \times 7 = 210$$
$$111 \times 7 = 777$$

I'd like to cash this $1,000 check. I'd like multiples of tens, please.

Common Multiples

Common multiples are multiples that are shared by two (or more) different numbers.

Some multiples of 8 are
8, 16, 24, 36, 40, 48, 56, 64.

Some multiples of 12 are
12, 24, 36, 48, 60, 72, 84.

Some **common multiples** of **8** and **12** are 24 and 48.

Least Common Multiples

The least common multiple (LCM) of two or more numbers is the common multiple (other than zero) with the least value.

Get Sharp Tip # 5
Zero (0) is a multiple of every number.

Some multiples of **5** are 5, 10, 15, 30, 35, 40, 45, 50, 85, 90.

Some multiples of **9** are 9, 18, 27, 36, 45, 54, 63, 72, 81, 90.

Some multiples of **15** are 15, 30, 45, 60, 75, 90.

Some **common multiples** of **5, 9,** and **15** are 45 and 90.

The **least common multiple** (LCM) is 45.

Get Sharp: Kinds of Numbers

Powers, Exponents, & Scientific Notation

Powers

The **power** of a number means the number of times the number is multiplied by itself.

Exponents

An **exponent** is a tiny number written to the right of the number and slightly above it. The exponent shows how many times the number has been multiplied by itself.

$2 = 2$	2 is the first power of 2 (or 2^1)
$2 \times 2 = 4$	4 is the second power of 2 (or 2^2)
$2 \times 2 \times 2 = 8$	8 is the third power of 2 (or 2^3)
$2 \times 2 \times 2 \times 2 = 16$	16 is the fourth power of 2 (or 2^4)
$2 \times 2 \times 2 \times 2 \times 2 = 32$	32 is the fifth power of 2 (or 2^5)

ASK

DR. CAL Q. LAYTER
He has a calculating mind!

How do I write a number in scientific notation?

Answer:

1. **Move the decimal point to the left far enough to form a number between 1 and 10.**

2. **Count the number of places you move the decimal point. Use this number as the exponent for writing a power of 10.**

Scientific Notation

Scientific notation is a way to use powers and exponents to write very large numbers. A number written in scientific notation is expressed as a product of a number between 1 and 10 and a power of the number 10.

Earth is **149,600,000** kilometers from the sun. A shorter way to write this number is:

1.496 x 10⁸

Pluto is **5,900,000,000** kilometers from the sun. A shorter way to write this is:

5.9 x 10⁹

Square Roots

Squares- A **square** is the name given to a number that has been raised to the second power (multiplied once by itself).

4 x 4 = 16. 16 is 4 squared.

Square Roots- A **square root** is the base number that was multiplied by itself once.

9 x 9 = 81. 81 is 9 squared. 9 is the square root of 81.

Radical Signs- A **radical sign** ($\sqrt{\ }$) is the symbol used to stand for square root.

$$\sqrt{49} = 7$$

Square Roots to 50
(Whole Numbers)

$\sqrt{1}$ = 1	$\sqrt{121}$ = 11	$\sqrt{441}$ = 21	$\sqrt{961}$ = 31	$\sqrt{1681}$ = 41
$\sqrt{4}$ = 2	$\sqrt{144}$ = 12	$\sqrt{484}$ = 22	$\sqrt{1024}$ = 32	$\sqrt{1764}$ = 42
$\sqrt{9}$ = 3	$\sqrt{169}$ = 13	$\sqrt{529}$ = 23	$\sqrt{1089}$ = 33	$\sqrt{1849}$ = 43
$\sqrt{16}$ = 4	$\sqrt{196}$ = 14	$\sqrt{576}$ = 24	$\sqrt{1156}$ = 34	$\sqrt{1936}$ = 44
$\sqrt{25}$ = 5	$\sqrt{225}$ = 15	$\sqrt{625}$ = 25	$\sqrt{1225}$ = 35	$\sqrt{2025}$ = 45
$\sqrt{36}$ = 6	$\sqrt{256}$ = 16	$\sqrt{676}$ = 26	$\sqrt{1296}$ = 36	$\sqrt{2116}$ = 46
$\sqrt{49}$ = 7	$\sqrt{289}$ = 17	$\sqrt{729}$ = 27	$\sqrt{1369}$ = 37	$\sqrt{2209}$ = 47
$\sqrt{64}$ = 8	$\sqrt{324}$ = 18	$\sqrt{784}$ = 28	$\sqrt{1444}$ = 38	$\sqrt{2304}$ = 48
$\sqrt{81}$ = 9	$\sqrt{361}$ = 19	$\sqrt{841}$ = 29	$\sqrt{1521}$ = 39	$\sqrt{2401}$ = 49
$\sqrt{100}$ = 10	$\sqrt{400}$ = 20	$\sqrt{900}$ = 30	$\sqrt{1600}$ = 40	$\sqrt{2500}$ = 50

Better Grades & Higher Test Scores / MATH
©Incentive Publications, Inc., Nashville, TN

Get Sharp: Kinds of Numbers

Integers

The set of **integers** includes all the positive numbers and all the negative numbers and zero (0).

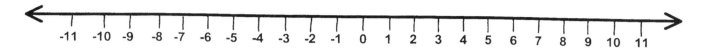

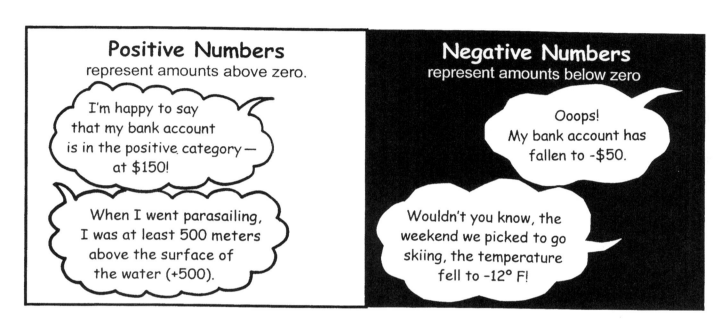

Positive Numbers
represent amounts above zero.

> I'm happy to say that my bank account is in the positive category — at $150!

> When I went parasailing, I was at least 500 meters above the surface of the water (+500).

Negative Numbers
represent amounts below zero

> Ooops! My bank account has fallen to -$50.

> Wouldn't you know, the weekend we picked to go skiing, the temperature fell to –12° F!

Opposites – **Opposites** are any two numbers that are the same distance from 0 but are on opposite sides of 0.

6 and – 6 are opposites   - 55¼ and 55¼ are opposites

100.7 and – 100.7 are opposites

Absolute Value – The **absolute value** of a number is its distance from 0 on a number line.

The absolute value of 500 is 500. The absolute value of –500 is the same: 500. The symbol |7| represents absolute value.

|88| = 88 |25½| = 25½ |-12.68| = -12.68

> Is the absolute value of 400 the same as - 400?

> Absolutely!

Fractions & Decimals

Fractions

Fraction means *part of a whole or part of a set*.
The set of **fractions** includes any number in the form of a/b that compares part of an object or part of a set to the whole. (The bottom number (b) cannot be 0.)

A fractional number has two parts: a numerator (top number) and a denominator (bottom number).

The **numerator** tells the number of parts that are being counted.

The **denominator** tells the number of parts in the whole.

(See more about fractions on pages 100-109 and 116-117.)

Drats! $\frac{7}{8}$ of my pizza has been eaten by $\frac{2}{3}$ of my friends!

Decimals

The set of **decimals** includes all numbers in a base ten system. The term *decimal,* however, is often used to describe numbers that use a decimal point to show an amount in between two whole numbers. (Many decimal numerals have digits in places to the right of the decimal point.)

Sample decimal numerals:

0.706   18.5   -36.07

-127.3   100.005

Place Value in Decimals

Learn these places.

tenths   hundredths   thousandths

5.12345

ten thousandths   hundred thousandths

5.1   reads *five and one tenth.*

5.12   reads *five and twelve hundredths.*

5.123   reads *five and one hundred twenty-three thousandths.*

5.1234   reads *five and one thousand two hundred thirty-four ten thousandths.*

5.12345   reads *five and twelve thousand three hundred forty-five hundred thousandths.*

See more about decimals and place value on pages 110-120.

Rational & Irrational Numbers

Rational Numbers

The **set of rational numbers** includes any number that can be written as a ratio (a/b) where a and b are both integers and where neither is zero. All terminating and repeating decimals are rational numbers.

$$\frac{3}{4} \qquad 0.25 \text{ or } \frac{1}{2}$$

$$12 \left(\text{or } \frac{12}{1}\right) \qquad -103 \left(\text{or } -\frac{103}{1}\right)$$

$$-\frac{9}{11} \qquad \frac{7}{8}$$

$$-12\frac{1}{2} \qquad 155\frac{3}{4} \qquad -.087 \qquad 69.33$$

Irrational Numbers

The **set of irrational numbers** includes any number that does not terminate and is nonrepeating.

$$0.454454445 \ldots \qquad -43.1428753 \ldots$$

$$1000.01101201 \ldots$$

Is a rational number a number that behaves reasonably or thinks clearly?

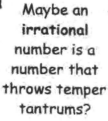

Maybe an irrational number is a number that throws temper tantrums?

Comparing & Ordering Numbers

Lots of problems and mathematical operations require you to compare numbers to each other. Any time you need to put numbers in order, you must compare the numbers and the amounts they represent. Compare numbers by paying careful attention to place value. Also, it is important to pay attention to signs for positive and negative numbers. (See more about comparing fractions, decimals, and integers on pages 102 and 111).

These are some special symbols for comparing numbers:

>   means *greater than*

<   means *less than*

=   means *equal to*

10,000	<	100,000
565	<	656
10,000	=	10^4
25½	>	24¼
75.67	<	76.057
¾	>	½
½	=	0.5
15^4	>	15^3
20^3	=	8,000

My biceps measure 25 cm.

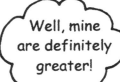

Well, mine are definitely greater!

Look at this group of numbers. It is out of order.

$\frac{1}{2}$

0.125

41,000

$99\frac{3}{4}$

$\sqrt{625}$

505,005,050

0.0125

10

505,005,005

Compare each number to every other number. Here is the same group of numbers in correct order.

0.0125

0.125

$\frac{1}{2}$

10

$\sqrt{625}$

$99\frac{3}{4}$

41,000

505,005,050

505,005,005

Patterns

Many times, sets of numbers follow patterns.
Here are a few:

2560, 1280, 640, 320, 160, 80, 40, 20, 10. . .
(Each number is half of the previous number.)

2; 4; 16; 256; 65,536; . . .
(Each number is the square of the previous number.)

Fibonacci Numbers

An Italian mathematician named Leonardo Fibonacci noticed patterns in nature. He described the patterns with a sequence of numbers:

1, 2, 3, 5, 8, 13, 21, 34, 55, 89, 144, 233, 377, 610. . .

(Each number except one is the sum of the two numbers to the left.)

Palindromic Numbers

Palindromic numbers follow a pattern that is quite fun!

These numbers read the same forwards and backwards!

8668

11,011

976,525,679

Here's a special trick that always produces a palindrome!

Sometimes you will need to continue the process longer to get a palindrome.

Try it with any number!

Choose ANY number. — 685

Reverse the digits to make a new number. — 586

Add the two numbers. — 685 + 586 = 1,271

Take the sum and reverse the digits. Add the two numbers. — WOW! — 1,271 + 1,721 = 3,993 — You get a palindrome every time!

GET SHARP →

on

Operations with Whole Numbers

Addition

Addition is the combining of two or more numbers or amounts.

$$123 + 300 + 1{,}000 + 255 = 1{,}6478$$

addends — sum

$$
\begin{array}{r}
4 \\
+\,5 \\
\hline
9
\end{array}
\qquad
\begin{array}{r}
623 \\
+\,174 \\
\hline
797
\end{array}
$$

addends — sum

The symbol for addition is

+

The word used for addition is **plus**.

The numbers being combined are **addends**.

The number resulting from an addition is a **sum**.

Each year in the U.S., about 5,100,000 people go to hospital emergency rooms with pains in the stomach. About 2,500,000 visit the emergency room with head pain. Another 4,500,000 go with chest pains.

I wonder how many of those head pains occur during math tests?

$$
\begin{array}{r}
4{,}500{,}000 \\
5{,}100{,}000 \\
+\,2{,}500{,}000 \\
\hline
12{,}100{,}000
\end{array}
$$

Better Grades & Higher Test Scores / MATH
©Incentive Publications, Inc., Nashville, TN

Addition with Carrying or Renaming

When addition results in a sum greater than 9 in any place, any amount over 10 is **carried** to the next place.

In the ones place of this example, **5 + 8 = 13.**

The 3 is written in the **ones** place, and the rest of the amount (1 ten) is **carried** over to the **tens** place. That extra ten is added to the other tens in the column. **Renaming** and **regrouping** are other terms used for this process.

In the example on the right, 9 + 6 equals 15.
Since these digits are in the tens place, the sum of these digits has a value of 15 tens.

The amount of 15 tens is **renamed** as **5 tens and 1 hundred**.
The 5 is written in the **tens** place, and the 1 (value of 100) is added to the other amounts in the **hundreds** column.

Several addends lined up beneath each other from a column
This is called *column addition.*

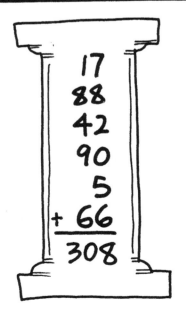

Subtraction

Subtraction is the operation of finding a missing addend (or, the taking away of one number or amount from another).

$$5{,}505{,}000 - 5{,}000{,}000 = 505{,}000$$

minuend   subtrahend   difference

The symbol for subtraction is

–

The word used for subtraction is **minus**.

The number being subtracted from is the **minuend**.

The number being subtracted is the **subtrahend**.

9 ← minuend → **847**		
- 3 ← subtrahend → **- 237**		
6 ← difference → **610**		

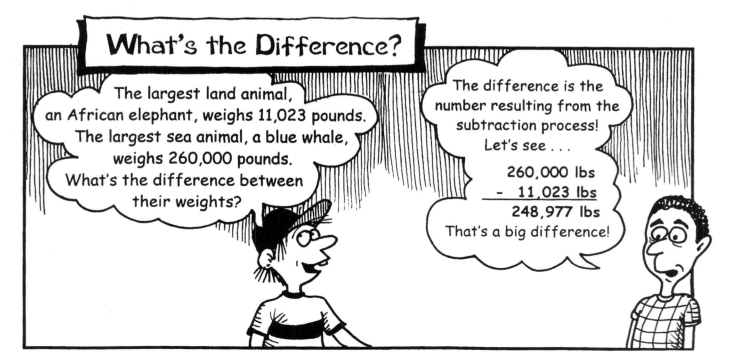

What's the Difference?

The largest land animal, an African elephant, weighs 11,023 pounds. The largest sea animal, a blue whale, weighs 260,000 pounds. What's the difference between their weights?

The difference is the number resulting from the subtraction process! Let's see . . .

```
  260,000 lbs
-  11,023 lbs
  248,977 lbs
```
That's a big difference!

Get Sharp: Operations

Better Grades & Higher Test Scores / MATH
©Incentive Publications, Inc., Nashville, TN

Subtraction with Borrowing (or Renaming)

Sometimes a digit in the minuend is smaller than the digit of the same place in the subtrahend. When this happens, it is necessary to **borrow** from the next place to the left.

Borrowing is the same as **renaming**. It means exchanging a ten to make a number in the ones place **larger** than the digit in the subtrahend. Or, it might mean exchanging a hundred for 10 tens, or a thousand for 10 hundreds etc.

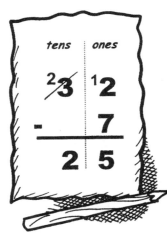

In this example, 7 is too large to subtract from 2. So, one of the tens is **borrowed** *or* **renamed** as 10 ones. Now there are 12 ones and 7 can be subtracted from 12. That leaves only 2 tens. (See the 2 written above the tens place.)

In this, the 6 in the tens place is smaller than the 9 in the ones place. So one of the hundreds is **borrowed** *or* **renamed** as 10 tens. Now there are 16 tens and 9 can be easily subtracted from 16. This leaves only 6 hundreds. (See the 6 written above the hundreds place.)

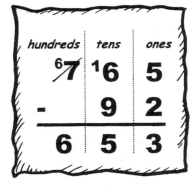

Addition & Subtraction Are Relatives!

Addition and subtraction are opposite (inverse) operations.

9 + 7 = 16
16 − 9 = 7
and 16 − 7 = 9

An addition problem can be checked with subtraction.

8,246	17,373
+ 9,127	− 9,127
17,373	8,246

A subtraction problem can be checked with addition.

50,000	49,536
− 464	+ 464
49,536	50,000

Get Sharp Tip #8

Renaming is also called regrouping.

79

Multiplication

Multiplication is repeated addition. When you multiply, you are adding the same number over and over again.

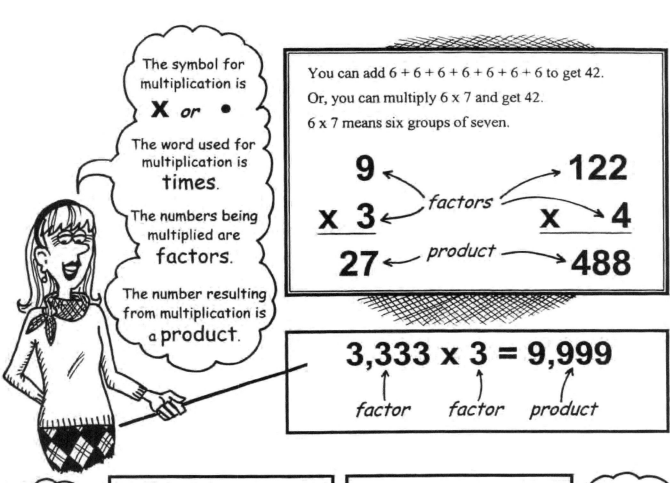

The symbol for multiplication is
X *or* **•**

The word used for multiplication is **times**.

The numbers being multiplied are **factors**.

The number resulting from multiplication is a **product**.

You can add $6 + 6 + 6 + 6 + 6 + 6 + 6$ to get 42.

Or, you can multiply 6×7 and get 42.

6×7 means six groups of seven.

$$
\begin{array}{r} 9 \\ \times\ 3 \\ \hline 27 \end{array}
$$
factors
product

$$
\begin{array}{r} 122 \\ \times\ 4 \\ \hline 488 \end{array}
$$

$$3{,}333 \times 3 = 9{,}999$$

factor　　*factor*　　*product*

Multiplying a number by one changes nothing!

Multiplying by One

ANY number multiplied by one has a product the same as the number!

$65 \times 1 = 65$

$999{,}999 \times 1 = 999{,}999$

Multiplying by Zero

ANY number multiplied by zero is **O**!

$65 \times 0 = 0$

No matter HOW BIG the number is, the product is still ZERO.

$0 \times 999{,}999{,}999{,}999 = 0$

Multiplying a number by zero gets you nothing!

Multiplication with Renaming

Sometimes you will need to rename (or regroup) numbers to complete a multiplication task. Here's how it works.

Step 1: Multiply the ones. 6 x 9 = 54 ones.
Rename the 54 ones as 5 tens and 4 ones.
Write the 5 above the tens column and the 4 in the ones place in the product.

Step 2: Multiply the tens: 6 x 3 = 18 tens.
Add the 5 tens. 18 + 5 = 23 tens.
Rename the 23 tens as 2 hundreds and 3 tens.
Write the 2 above the hundreds column and the 3 in the tens place in the product.

Step 3: Multiply the hundreds: 6 x 8 = 48 hundreds.
Add the 2 hundreds: 48 + 2 = 50 hundreds
Rename the 50 hundreds as 5 thousands and 0 hundreds.
Write the 0 in the hundreds place and the 5 in the thousands place in the product.

Multiplication with Multiples of Ten

6 x 300 =

Step 1:
Drop the zeros and rewrite the problem as a basic multiplication fact.

6 x 3

Step 2:
Find the product:

6 x 3 = 18

Step 3:
At the end of the product, write the same number of zeros you dropped.

1,800

More Examples:

20,000 x 140
2 x 14 = 28
28 *with* 5 zeros = 2,800,000

80 x 5,000,000
8 x 5 = 40
40 *with* 7 (more) zeros = 400,000,000

Better Grades & Higher Test Scores / MATH
©Incentive Publications, Inc., Nashville, TN

Get Sharp: Operations

Multiplication by Larger Numbers

ten thousands	thousands	hundreds	tens	ones
		3	6	8
x		2	5	7
	2	5	7	6
1	8	4	0	
+ 7	3	6		
9	4,	5	7	6

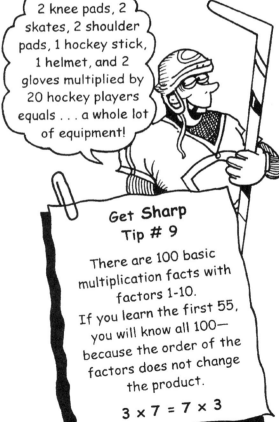

2 knee pads, 2 skates, 2 shoulder pads, 1 hockey stick, 1 helmet, and 2 gloves multiplied by 20 hockey players equals . . . a whole lot of equipment!

Get Sharp Tip # 9

There are 100 basic multiplication facts with factors 1-10.
If you learn the first 55, you will know all 100— because the order of the factors does not change the product.

$3 \times 7 = 7 \times 3$

Step 1: Multiply by ones.

Multiply 7 x 8 (7 x 8 = 56 ones).
Rename the 56 ones as 5 tens and 6 ones.

Multiply 7 x 6 (7 x 6 = 42 tens).
Add the 5 tens. (42 + 5 = 47 tens)
Rename the 47 tens as 4 hundreds and 7 tens.

Multiply 7 x 3 (7 x 3 = 21 hundreds).
Add the 4 hundreds. (21 + 4 = 25 hundreds)
Rename the 25 hundreds as 2 thousands
    and 5 hundreds.

Step 2: Multiply by tens.

Multiply 5 x 8 (5 x 8 = 40 tens)
Rename the 40 tens as 4 hundreds and 0 tens.

Multiply 5 x 6 (5 x 6 = 30 hundreds).
Add the 4 hundreds. (30 + 4 = 34 hundreds)
Rename the 34 hundreds as 3 thousands
    and 4 hundreds.

Multiply 5 x 3 (5 x 3 = 15 thousands).
Add the 3 thousands (15 + 3 = 18 thousands).
Rename the 18 thousands as 1 ten thousand
    and 8 thousands.

Step 3: Multiply by hundreds.

Multiply 2 x 8 (2 x 8 = 16 hundreds)
Rename the 16 hundreds as 1 thousand
    and 6 hundreds.

Multiply 2 x 6 (2 x 6 = 12 thousands).
Add the 1 thousand. (12 + 1 = 13 thousand)
Rename the 13 thousands as 1 ten thousand
    and 3 thousands.

Multiply 2 x 3 (2 x 3 = 6 ten thousands).
Add the 1 ten thousand. (6 + 1 = 7 ten thousands.)

Step 4: Add the columns.

Division

Division is repeated subtraction. When you divide,
you are subtracting the same number over and over again.

You can subtract 8 from 48 six times: $48 - 8 - 8 - 8 - 8 - 8 - 8 = 0$

Or, you can divide $48 \div 8$ and get 6.

$46 \div 8 = 6$ means there are 6 groups of 8 in 48.

The symbol for division is ⌐ or ÷

Division is a way of finding out how many times one number **(the divisor)** will fit into another number **(the dividend).**

The number resulting from division is the **quotient.**

If a divisor does not fit an even number of times into a dividend, there will be a number left over. This is called the **remainder.**

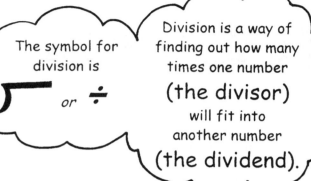

$$\overset{\text{quotient}}{8} \\ \text{divisor } 9\overline{)72} \leftarrow \text{dividend}$$

$$\underset{\text{dividend}}{9{,}633} \div \underset{\text{divisor}}{3} = \underset{\text{quotient}}{3{,}211}$$

A **fraction bar** also symbolizes division.

$$\frac{dividend}{divisor} = quotient \qquad \frac{125}{25} = 5$$

This week, my attention is equally divided between Mark Picante and Russell Largo.

$$\overset{7 \text{ R}2}{11\overline{)79}}$$
$$remainder$$
$$59 \div 7 = 8 \text{ R } 3$$

Better Grades & Higher Test Scores / MATH
©Incentive Publications, Inc., Nashville, TN

Get Sharp: Operations

Dividing a Number by Itself

ANY number divided by itself yields **1**!

65 ÷ 65 = 1

No matter HOW BIG the number is, the quotient is still ONE.

999,999,999 ÷ 999,999,999 = 1

Dividing a Number by One

ANY number divided by 1 yields that number.

95 ÷ 1 = 95

700,000 ÷ 1 = 700,000

Get Sharp Tip # 10

There is no division by zero. It's impossible!

Divisibility

A number is ***divisible*** by another number if the quotient of the two numbers is a whole number. (50 is divisible by 5 because the quotient is a whole number (10).

A number is *divisible by 2* if the last digit is 0, 2, 4, 6, or 8.

A number is *divisible by 3* if the sum of its digits is divisible by 3.

A number is *divisible by 4* if the last two digits are divisible by 4.

A number is *divisible by 5* if the last digit is 0 or 5.

A number is *divisible by 6* if the number is divisible by both 2 and 3.

A number is *divisible by 8* if the last three digits are divisible by 8.

A number is *divisible by 9* if the sum of its digits is divisible by 9.

A number is *divisible by 10* if the last digit is 0.

Multiplication & Division Are Relatives!

Multiplication and division are opposite (inverse) operations.

$9 \times 7 = 63$
$63 \div 9 = 7$, and
$63 \div 7 = 9$

A multiplication problem can be checked with division.

$$\begin{array}{r} 8,246 \\ \times 5 \\ \hline 41,230 \end{array}$$

$41,230 \div 5 = 8,246$

A division problem can be checked with multiplication.

$4,869 \div 9 = 541$
$541 \times 9 = 4,869$

84

Better Grades & Higher Test Scores / MATH
©Incentive Publications, Inc., Nashville, TN

Division With One-Digit Divisors

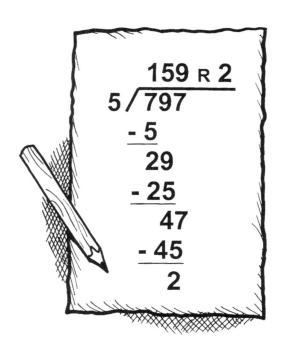

$$159 \text{ R } 2$$
$$5\overline{)797}$$
$$-5$$
$$29$$
$$-25$$
$$47$$
$$-45$$
$$2$$

Step 1: Does 5 go into 7? (yes-1 time)

Write the 1 above the 7.
Multiply 1 x 5. Write the product under the 7.
Subtract 7 – 5 (=2).
Bring the next digit (9) down next to the 2.

Step 2: Does 5 go into 29? (yes-5 times)

Write the 5 above the 9.
Multiply 5 x 5. Write the product under 29.
Subtract 29-25 (=4).
Bring the next digit (7) down next to the 4.

Step 3: Does 5 go into 47? (yes-9 times)

Write the 9 above the 5.
Multiply 9 x 5. Write the product under 47.
Subtract 47 - 45 (=2).
Write the remainder (2) next to the quotient.

— Division with Multiples of Ten —

$$72,000 \div 100 =$$

Step 1:
Place a decimal point after the dividend:

72,000.

Step 2:
Move the decimal point one place to the left for each zero in the divisor:

720.00

(The decimal point was moved 2 places because 100 has 2 zeros.)

Step 3:
Drop any zeros to the right of the decimal point:

72,000 ÷ 100 = 720

More Examples:

123,000,000 ÷ 1,000
123,000,000.
123,000.000

123,000,000 ÷ 1,000 = 123,000

800,000 ÷ 10,000
800,000.
80.0000

800,000 ÷ 10,000 = 80

Better Grades & Higher Test Scores / MATH
©Incentive Publications, Inc., Nashville, TN

Get Sharp: Operations

Division with Larger Divisors

When the divisor has more than one digit, division problems can get very tricky.
Here are some steps to help you handle this process without feeling baffled.

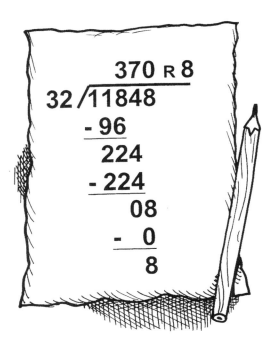

Step 1: Does 32 go into 1? (no).

Step 2: Does 32 go into 11? (no)

Step 3: Does 32 go into 118? (yes)

>Round 32 to the closest 10. (30)
>Estimate the number of 30s in 118. *(about 3)*
>Write 3 above the 8 of 118.
>Multiply 3 x 32. Write the product under 118.
>Subtract 118 - 96 (= 22).
>Bring the next digit (4) down next to the 22.

Step 4: Does 32 go into 224? (yes)

>Round 32 to 30 again.
>Estimate the number of 30s in 224. *(about 7)*
>Write 7 above the 4 of 1184.
>Multiply 7 x 32. Write the product under the 224.
>Subtract 224 – 224 (- 0).
>Bring down the next digit (8) next to the 0.

Step 5: Does 34 go into 8? (no)

>Write 0 above the 8 of 11848.
>Multiply 0 x 32. Write the product under 8.
>Subtract 8 – 0 (-8).
>8 is smaller than the divisor, 32.
>Therefore, 8 is the remainder.
>Write the remainder next to the quotient.

I just read that an elephant's brain weighs about 6000 grams. That's 200 times the weight of a cat's brain. 6000 grams divided by 200 equals 30 grams. Aha! A cat's brain weighs 30 grams!

Trivia Facts

I wonder, can an elephant do long division faster than a cat?

Here's a better question: Will the elephant do my math homework for me?

Averages

An **average** is the sum of many numbers or amounts divided by the number of numbers!

How to find an average:

To find the average of several numbers, add them together.
Then divide the sum by the number of items (or numbers).

The average of these numbers is 61.

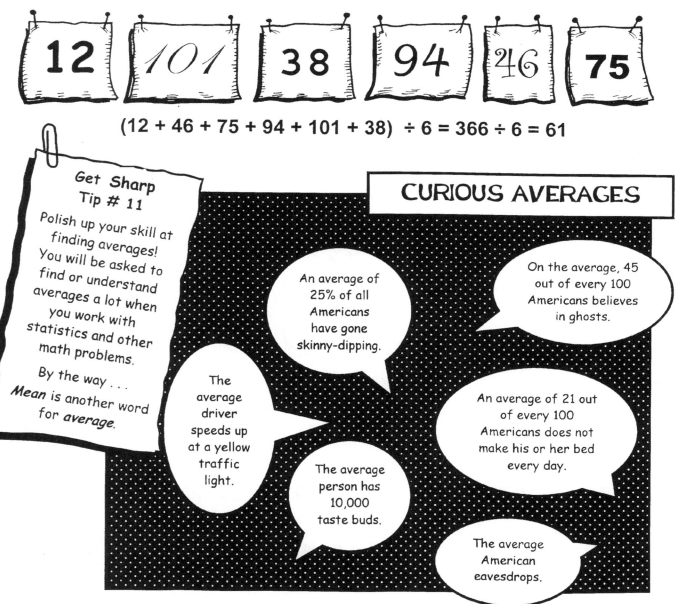

12 | **101** | **38** | **94** | **46** | **75**

$$(12 + 46 + 75 + 94 + 101 + 38) \div 6 = 366 \div 6 = 61$$

Get Sharp Tip # 11

Polish up your skill at finding averages! You will be asked to find or understand averages a lot when you work with statistics and other math problems.

By the way . . .

Mean is another word for *average*.

CURIOUS AVERAGES

An average of 25% of all Americans have gone skinny-dipping.

On the average, 45 out of every 100 Americans believes in ghosts.

The average driver speeds up at a yellow traffic light.

The average person has 10,000 taste buds.

An average of 21 out of every 100 Americans does not make his or her bed every day.

The average American eavesdrops.

Better Grades & Higher Test Scores / MATH
©Incentive Publications, Inc., Nashville, TN

Get Sharp: Operations

Properties

Properties for Addition

Commutative Property

The order in which numbers are added does not affect the sum.

Examples: **6 + 4 = 4 + 6**   *(sum is 10 no matter what the order)*

50,000 + 5, 000 + 500 = 500 + 5,000 + 50,000

1,500 + 33 + 20,000 = 20,000 + 33 + 60 + 1500

Associative Property

The way in which numbers are grouped does not affect the sum.

Examples: **7 + (3+2) = (7+3) + 2** *(sum is 12 no matter what the grouping)*

60 + (100 + 5) + (85 + 9) = (60 + 100) + (5 + 85 + 9)

Identity Property
(or Zero Property)

The sum of 0 and any number is that number.

Examples: **7 + 0 = 7**

486 + 0 = 486

9,000,000,000,000 + 0 = 9,000,000,000,000

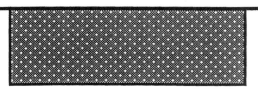

Better Grades & Higher Test Scores / MATH
©Incentive Publications, Inc., Nashville, TN

Properties for Multiplication

Commutative Property

The order in which numbers are multiplied does not affect the product.

Example: $6 \times 9 = 9 \times 6$ *(product is 56 no matter what the order)*

Associative Property

The way in which numbers are grouped does not affect the product.

Example: $(8 \times 3) \times 4 = 8 \times (3 \times 4)$ *(product is 96 no matter what the grouping)*

Identity Property
(or Property of One)

The product of 1 and any number that is that number.

Example: $99 \times 1 = 99$ $50,000 \times 1 = 50,000$

Zero Property

The product of zero and any number is zero.

Examples: $750 \times 0 = 0$ $0 \times 6,543 = 0$

Distributive Property
(of Multiplication Over Addition)

To multiply a sum of numbers, first add the numbers in parentheses and then multiply the sum.

Example: $10 \times (50 + 5) = 10 \times 55 = 550$

OR Multiply the numbers separately and then add the products.

Example: $10 \times (50 + 5) = (10 \times 50) + (5 \times 50) = 500 + 50 = 550$

Better Grades & Higher Test Scores / MATH
©Incentive Publications, Inc., Nashville, TN

Get Sharp: Operations

Which Operation?

Be Sharp About Operations!

Sometimes you'll be asked to find the missing operation in an equation. In word problems you always have to decide the correct operation to use. These hints will help you decide which operation is correct.

1. If the answer is larger than the other numbers, the operation is probably ***addition or multiplication.***

55 ☐ 42 = 97

2. If the answer is smaller than the other numbers, the operation is probably ***subtraction or division.***

600,000,000 ☐ 2,000 = 300,000

5,609 ☐ 832 = 4,777

3. If the answer is quite a bit larger than either number, the operation is probably ***multiplication.***

55 ☐ 42 = 2310

4. If the equation has a fraction, and no other numbers, it's likely there's some ***division*** to be done

$$\frac{7}{10} = x$$

My grandma's operation was a marvel of mathematical genius. Three teams of doctors performed seven separate procedures in precisely five hours.

And . . . Grandma's facelift, nose job, hair transplant, tummy tuck, teeth implants, chin tuck, and liposuction look great!

5. Watch for these words in problems or directions. They are signals for ***addition!***

sum	*total*	*all together*
together	*both*	*increased by*

6. Watch for these words in problems or directions. They are signals for ***multiplication!***

times	*how many times*
a product of	*twice as much as*

7. Watch for these words in problems or directions. They are signals for ***subtraction!***

difference	*less than*	*left over*
remain	*take away*	*have left*
fewer than	*much more*	*much less*

8. Watch for these words in problems or directions. They are signals for ***division!***

divided by	*average*	*half as much*
any fraction	*sharing*	*equal parts*

Better Grades & Higher Test Scores / MATH
©Incentive Publications, Inc., Nashville, TN

Some Interesting Number Tricks

The 3-Digit Trick

The final answer will always be the original number.

Step 1

Write any three digits.

472

Step 2

Repeat the numbers in the same order to get a six-digit number.

472,472

Step 3

Divide the six-digit number by 13.

$$13\overline{)472{,}472} = 36{,}344$$

Step 4

Divide the resulting number by 11.

$$11\overline{)36{,}344} = 3{,}304$$

Step 5

Divide the resulting number by seven.

$$7\overline{)3{,}304} = 472$$

Step 6

The final answer will always be the original number.

472 !

Goldbach's Guesses

In 1742, a man named Christian Goldbach made these two "guesses" (theories). No one has ever been able to prove that either is *always* true, but no one has ever found a case in which either is false.

Experiment with them on your own!

Every even number greater than 2 is the sum of two prime numbers.
6 = 1 + 5
162 = 79 + 83

Every odd number greater than 7 is the sum of three prime numbers.
17 = 5 + 5 + 7
139 = 73 + 29 + 37

The Reappearing Number Trick

Choose any four digits and follow these directions. The same number will appear every time.

1. Choose four different digits from 0 to 9.

1,572

2. Arrange the numbers to make the largest possible number. Arrange the numbers again to make the smallest possible number. Subtract the smaller number from the larger number.

```
  7,521
- 1,257
  6,264
```

3. Arrange the individual numbers in the resulting answer to make the largest and smallest possible numbers. Subtract the numbers as before.

```
  6,642
- 2,466
  4,176
```

4. Continue the process in the same manner. Eventually you will get to the number **6174**, no matter what four numbers you choose.

```
  7,641
- 1,467
  6,174
```

A Multiplication Trick

Here's a fun way to check multiplication problems.

Step 1. First, work the problem and write the answer.

Step 2. Then add the digits in each factor.

Step 3. Now add the digits in each answer of step 2.

Step 4. Multiply the two answers from step 3.

Step 5. Add the digits in the product. Continue adding the digits in the resulting answers until you get a one-digit answer.

Step 6. Now add the digits in the product of the original problem.

Step 7. Continue adding the digits in the resulting answers until you get a one-digit answer.

Step 8. Compare answers from steps 5 and 7. If both 1-digit answers are the same, the product of the original answer is correct. If the problem was not done correctly to begin with, these last two numbers will not match.

The original problem:

$3{,}752 \times 1608 = 6{,}033{,}216$

$3 + 7 + 5 + 2 = 17; \quad 1 + 6 + 0 + 8 = 15$

$1 + 7 = 8; \quad 1 + 5 = 6$

$8 \times 6 = 48$

$4 + 8 = 12; \quad 1 = 2 = 3$

$6 + 0 + 3 + 3 + 2 + 1 + 6 = 21$

$2 + 1 = 3$

Better Grades & Higher Test Scores / MATH
©Incentive Publications, Inc., Nashville, TN

To find **an addition fact:**

Locate one addend on the top horizontal row. Locate the second addend on the left-most vertical row. Draw an imaginary line from each number to the box in which the lines meet. The number in that box is the sum of the two addends.

To find **a subtraction fact:**

Locate one addend in the top horizontal row. Draw an imaginary line from that number down the page vertically until you come to the number from which you are subtracting. Draw an imaginary line from that number to the left-most box. The number in that box is the missing addend.

+	1	2	3	4	5	6	7	8	9	10	11	12	13	14	15
1	2	3	4	5	6	7	8	9	10	11	12	13	14	15	16
2	3	4	5	6	7	8	9	10	11	12	13	14	15	16	17
3	4	5	6	7	8	9	10	11	12	13	14	15	16	17	18
4	5	6	7	8	9	10	11	12	13	14	15	16	17	18	19
5	6	7	8	9	10	11	12	13	14	15	16	17	18	19	20
6	7	8	9	10	11	12	13	14	15	16	17	18	19	20	21
7	8	9	10	11	12	13	14	15	16	17	18	19	20	21	22
8	9	10	11	12	13	14	15	16	17	18	19	20	21	22	23
9	10	11	12	13	14	15	16	17	18	19	20	21	22	23	24
10	11	12	13	14	15	16	17	18	19	20	21	22	23	24	25
11	12	13	14	15	16	17	18	19	20	21	22	23	24	25	26
12	13	14	15	16	17	18	19	20	21	22	23	24	25	26	27
13	14	15	16	17	18	19	20	21	22	23	24	25	26	27	28
14	15	16	17	18	19	20	21	22	23	24	25	26	27	28	29
15	16	17	18	19	20	21	22	23	24	25	26	27	28	29	30

Better Grades & Higher Test Scores / MATH
©Incentive Publications, Inc., Nashville, TN

Get Sharp: Operations

Multiplication & Division Facts Matrix

To find a multiplication fact:

Locate one factor on the top horizontal row and the second on the left-most vertical row. Draw an imaginary line from each number to the box in which the lines meet. The number in that box is the product of the two factors.

To find a division fact:

Locate the divisor on the top horizontal row. Draw an imaginary line from that number down the page vertically until you come to the number being divided. Draw an imaginary line from that number to the left-most box. The number in that box is the missing factor.

X	1	2	3	4	5	6	7	8	9	10	11	12	13	14	15
1	1	2	3	4	5	6	7	8	9	10	11	12	13	14	15
2	2	4	6	8	10	12	14	16	18	20	22	24	26	28	30
3	3	6	9	12	15	18	21	24	27	30	33	36	39	42	45
4	4	8	12	16	20	24	28	32	36	40	44	48	52	56	60
5	5	10	15	20	25	30	35	40	45	50	55	60	65	70	75
6	6	12	18	24	30	36	42	48	54	60	66	72	78	84	90
7	7	14	21	28	35	42	49	56	63	70	77	84	91	98	105
8	8	16	24	32	40	48	56	64	72	80	88	96	104	112	120
9	9	18	27	36	45	54	63	72	81	90	99	108	117	126	135
10	10	20	30	40	50	60	70	80	90	100	110	120	130	140	150
11	11	22	33	44	55	66	77	88	99	110	121	132	143	154	165
12	12	24	36	48	60	72	84	96	108	120	132	144	156	168	180
13	13	26	39	52	65	78	91	104	117	130	143	156	169	182	195
14	14	28	42	56	70	84	98	112	126	140	154	168	182	196	210
15	15	30	45	60	75	90	105	120	135	150	165	180	195	210	225

Get Sharp: Operations

Better Grades & Higher Test Scores / MATH
©Incentive Publications, Inc., Nashville, TN

GET SHARP →

on

Fractions & Decimals

I am a fraction of my former self.

Fractions

A **fraction** is any number written in the form of $\dfrac{a}{b}$

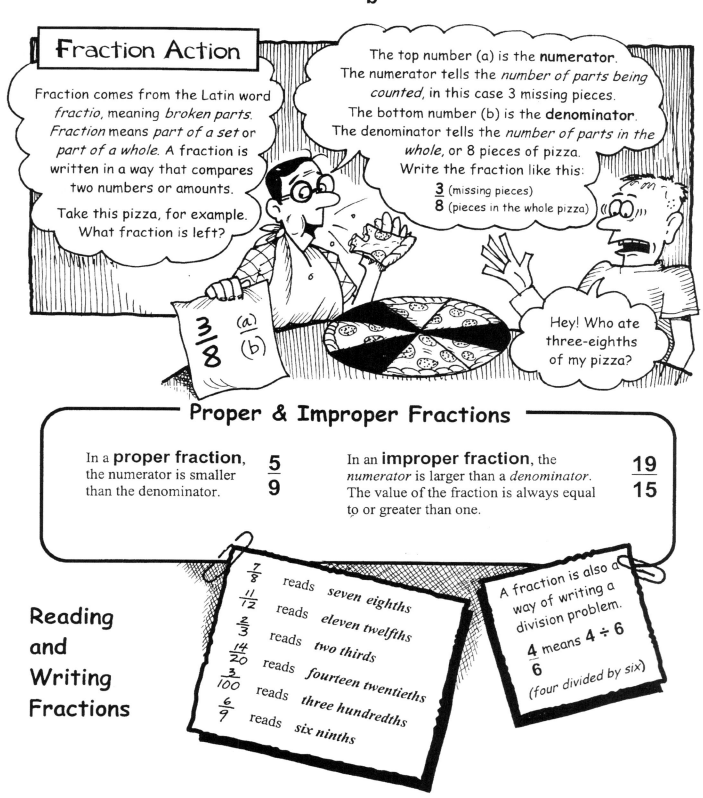

Fraction Action

Fraction comes from the Latin word *fractio,* meaning *broken parts.* *Fraction* means *part of a set* or *part of a whole.* A fraction is written in a way that compares two numbers or amounts.

Take this pizza, for example. What fraction is left?

The top number (a) is the **numerator**. The numerator tells the *number of parts being counted,* in this case 3 missing pieces.

The bottom number (b) is the **denominator**. The denominator tells the *number of parts in the whole,* or 8 pieces of pizza.

Write the fraction like this:

$\dfrac{3}{8}$ (missing pieces)
(pieces in the whole pizza)

$\dfrac{3}{8}$ (a)
(b)

Hey! Who ate three-eighths of my pizza?

Proper & Improper Fractions

In a **proper fraction**, the numerator is smaller than the denominator.

$\dfrac{5}{9}$

In an **improper fraction**, the *numerator* is larger than a *denominator.* The value of the fraction is always equal to or greater than one.

$\dfrac{19}{15}$

Reading and Writing Fractions

$\dfrac{7}{8}$ reads *seven eighths*

$\dfrac{11}{12}$ reads *eleven twelfths*

$\dfrac{2}{3}$ reads *two thirds*

$\dfrac{14}{20}$ reads *fourteen twentieths*

$\dfrac{3}{100}$ reads *three hundredths*

$\dfrac{6}{9}$ reads *six ninths*

A fraction is also a way of writing a division problem.

$\dfrac{4}{6}$ means $4 \div 6$

(four divided by six)

Fractions as Part of a Whole

Some fractions represent parts of a whole.
Each pizza is a whole item, cut into parts.

$\frac{5}{8}$ of the pizza is topped with anchovies.

$\frac{2}{8}$ of the pizza is topped with snails.

$\frac{1}{8}$ has both toppings!

$\frac{3}{10}$ of this pizza has been eaten!

Fractions as Part of a Set

Sometimes, a fraction represents part of a set. This is a set of friends.
The picture also includes a set of hats, a set of shoes, a set of noses,
a set of legs, a set of feet, and other sets.

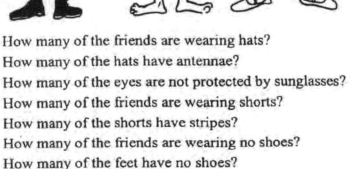

How many of the friends are wearing hats?	3/5	(the set: friends)
How many of the hats have antennae?	1/3	(the set: hats)
How many of the eyes are not protected by sunglasses?	8/10	(the set: eyes)
How many of the friends are wearing shorts?	4/5	(the set: friends)
How many of the shorts have stripes?	1/4	(the set: shorts)
How many of the friends are wearing no shoes?	1/5	(the set: friends)
How many of the feet have no shoes?	7/10	(the set: feet)
How many of the shoes are black?	2/7	(the set: shoes)

Better Grades & Higher Test Scores / MATH
©Incentive Publications, Inc., Nashville, TN

Get Sharp: Fractions

Mixed Fractional Numbers

A **mixed fractional number** combines a whole number
The value of a mixed fractional number is always greater than one
(unless it is a negative number).

I've already used $2\frac{1}{2}$ cans of paint
on this wall. It will take at least $4\frac{3}{4}$ more
to finish the job. Oh, oh!

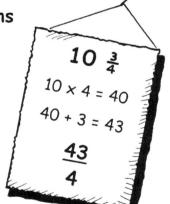

A mixed fractional number may be written as an improper fraction.
Sometimes it is useful to changed a mixed number into an improper
fraction in order to complete an operation.

How to Change Mixed Numbers to Improper Fractions

Step 1: Multiply the whole number by the denominator of the fraction.

Step 2: Add the numerator to the product of Step 1.

Step 3: Write the sum from Step 2 as the numerator for the new fraction.

Step 4: Write the original denominator as the denominator for the new fraction.

$10\frac{3}{4}$

$10 \times 4 = 40$

$40 + 3 = 43$

$\dfrac{43}{4}$

How to Change Improper Fractions to Mixed Numbers

$\dfrac{39}{2}$

$39 \div 2 = 19 \text{ R } 1$

$19\frac{1}{2}$

Step 1: Divide the numerator by the denominator.

Step 2: Write the quotient as a whole number.

Step 3: Write the remainder as the numerator in a fraction.

Step 4: Write the original denominator in the fraction.

Equivalent Fractions

Equivalent fractions are two or more fractions that represent the same amount.

J.J. left $\frac{2}{3}$ of his chocolate bar on the table.

Arty left $\frac{6}{9}$ of his chocolate bar on the table.

Roberta left $\frac{4}{6}$ of her chocolate bar on the table.

If the chocolate bars were the same size to begin with, all three friends have the same amount.

$\frac{2}{3}$ and $\frac{6}{9}$ and $\frac{4}{6}$ are equivalent. Write: $\frac{2}{3} = \frac{6}{9}$

How to Form Equivalent Fractions

Step 1: Multiply or divide both the numerator and the denominator by the same nonzero number.

Step 2: Write the new fraction.

$$\frac{3}{4} = \frac{3 \times 2}{4 \times 2} = \frac{6}{8}$$

$$\frac{56}{72} = \frac{56 \div 8}{72 \div 8} = \frac{7}{9}$$

How to Tell Equivalent Fractions

Step 1:

Cross multiply.

Step 2:

Compare the two products.

Step 3:

If the products are equal, the fractions are equivalent. Otherwise they are not.

$\frac{6}{15} \times \frac{4}{10}$ → $6 \times 10 = 60$ / $15 \times 4 = 60$ → $60 = 60$, *so the fractions are equivalent*

$\frac{7}{9} \times \frac{4}{5}$ → $7 \times 5 = 35$ / $9 \times 4 = 36$ → $35 = 36$, *so the fractions are not equivalent*

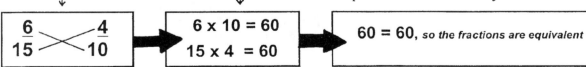

99

Fractions in Lowest Terms

A fraction is **in lowest terms** when the greatest common factor (GCF) of the numerator and the denominator is 1. *(See page 66 for an explanation of factors, common factors, and greatest common factors.)*

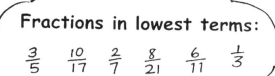

Fractions in lowest terms:

$$\frac{3}{5} \quad \frac{10}{17} \quad \frac{2}{7} \quad \frac{8}{21} \quad \frac{6}{11} \quad \frac{1}{3}$$

Fractions NOT in lowest terms:

$$\frac{22}{10} \quad \frac{4}{6} \quad \frac{3}{9} \quad \frac{16}{56} \quad \frac{7}{21} \quad \frac{5}{10}$$

How to Find the GCF of Two or More Numbers

Step 1:

Write the factors for each number.

4 and 16

The factors of **4** are **1, 2, 4**.
The factors of **16** are:
1, 2, 4, 8, and **16**.

Step 2:

Find the factors that are common to both numbers.

1, 2, 4
are common factors of **4** and **16**.

Step 3:

Find the greatest of these common factors.

The **GCF** of **4** and **16** is **4**.

How to Reduce a Fraction to Lowest Terms

Step 1: Find the greatest common factor (GCF) of the numerator and denominator.

$$\frac{24}{30}$$ The GCF is **6**.

Step 2: Divide the numerator and the denominator by that GCF.

$$\frac{24 \div 6 = 4}{30 \div 6 = 5}$$

Step 3: Write the quotients as a new numerator and denominator.

$$\frac{24}{30}$$ *in lowest terms is* $$\frac{4}{5}$$

Like & Unlike Fractions

Like Fractions

are fractions with the same denominator.

$$\frac{6}{8} \qquad \frac{2}{8}$$

$$\frac{7}{8} \qquad \frac{4}{8} \qquad \frac{1}{8} \qquad \frac{11}{8}$$

Unlike Fractions

have different denominators.

$$\frac{3}{10} \qquad \frac{75}{100} \qquad \frac{1}{15} \qquad \frac{3}{4}$$

$$\frac{5}{8} \qquad \frac{6}{19}$$

I like fractions.

Get Sharp Tip # 13
To find the **least common denominator** for two fractions, you need to find the **LCM** *(least common multiple)* for both denominators. Write several multiples for each denominator. Find the common ones. Then look for the smallest number common to both.

How do you **unlike** a fraction?

How to Change Unlike Fractions into Like Fractions

Step 1: Find the least common multiple (LCM) of both denominators. This will give you the least common denominator for the 2 fractions.

$$\frac{2}{3} \qquad \text{and} \qquad \frac{4}{5}$$

15 is the LCM for 3 and 5, so 15 is the least common denominator for the two fractions.

Step 2: Remember the number used as the multiplier to get the LCM for the first fraction. Multiply the numerator of the first fraction by that number.

To get a denominator of 15 for the first fraction, 3 was multiplied by 5. So multiply 2 by 5 also.

$$\frac{2 \times 5 = 10}{3 \times 5 = 15} \quad so \quad \frac{2}{3} = \frac{10}{15}$$

Step 3: Remember the number used as the multiplier to get the LCM for the second fraction. Multiply the numerator of the second fraction by that number.

To get a denominator of 15 for the second fraction, 5 was multiplied by 3. So multiply 4 by 3 also.

$$\frac{4 \times 3 = 12}{5 \times 3 = 15} \quad so \quad \frac{4}{5} = \frac{12}{15}$$

101

Comparing & Ordering Fractions

Sometimes you can look at a group of fractions, and know for sure
that some are smaller or larger than others. Sometimes you may not be sure.
Then you need some methods for finding out exactly how they compare.

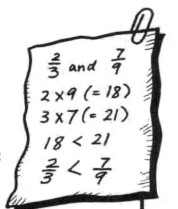

How to Compare & Order Two Fractions

To compare $\frac{2}{3}$ and $\frac{7}{9}$ cross multiply the two fractions.

Step 1: Multiply the first numerator and the second denominator: **2 x 9 = 18**

Step 2: Multiply the first denominator and the second numerator: **3 x 7 = 21**

Step 3: Compare products:

If the first multiplication has the greater product, the first fraction is greater.
If the second multiplication has the greater product, the second fraction is greater.

18 is less than 21, therefore $\frac{2}{3} < \frac{7}{9}$

How to Compare & Order Several Fractions

Change all the fractions to like fractions (common denominators).
Then it becomes easy to put them in order.

Get Sharp Tip # 14

If you're comparing mixed fractional numbers, first change them into improper fractions. Then follow either of the methods for comparing.

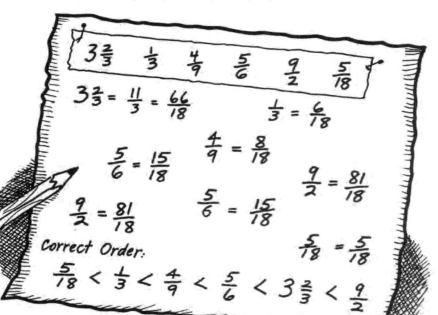

Adding & Subtracting Fractions

How to Add & Subtract Like Fractions

Step 1: If the fractions have like denominators, just add or subtract the numerators. (Denominators stay the same.)

Step 2: Reduce sums or differences to lowest terms.

$$\frac{5}{20} + \frac{3}{20} + \frac{7}{20} = \frac{15}{20} \xrightarrow{\textit{in lowest terms}} \frac{3}{4} \qquad \frac{8}{9} - \frac{5}{9} = \frac{3}{9} \xrightarrow{\textit{in lowest terms}} \frac{1}{3}$$

How to Add & Subtract Unlike Fractions

Step 1: Find the LCM for all denominators and change the fractions to like fractions.

Step 2: Add or subtract the numerators. (Denominators stay the same.)

Step 3: Reduce sums or differences to lowest terms.

$$\frac{1}{3} + \frac{2}{8} + \frac{1}{6} = \frac{8}{24} + \frac{6}{24} + \frac{4}{24} = \frac{18}{24} \xrightarrow{\textit{in lowest terms}} \frac{3}{4}$$

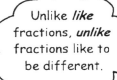

Unlike *like* fractions, *unlike* fractions like to be different.

How to Add & Subtract Mixed Numerals

Step 1: Change all mixed numerals to improper fractions.

Step 2: Find the LCM for all the denominators and change the fractions to like fractions.

Step 3: Add or subtract the numerators. (Denominators stay the same.)

Step 4: Reduce sums or differences to lowest terms.

$$19\frac{1}{2} - 8\frac{3}{4} = \frac{39}{2} - \frac{36}{4} = \frac{78}{4} - \frac{36}{4} = \frac{42}{4} = 10\frac{2}{4} = 10\frac{1}{2}$$

Better Grades & Higher Test Scores / MATH
©Incentive Publications, Inc., Nashville, TN

Get Sharp: Fractions

Multiplying Fractions

How to Multiply Fractions

Step 1: Multiply the numerators; the product is a new numerator.

Step 2: Multiply the denominators; the product is a new denominator.

Step 3: Reduce the product fraction to lowest terms.

$$\frac{5}{9} \times \frac{3}{11} = \frac{15}{99} \xrightarrow{\text{in lowest terms}} \frac{5}{33}$$

The top running speed of a human is less than $\frac{1}{2}$ the speed of an ostrich.

Only $\frac{1}{3}$ of all humans can flare their nostrils.

How to Multiply a Fraction by a Whole Number

Step 1: Multiply the numerator by the whole number.

Step 2: Write this product as the numerator in the answer.

Step 3: Write the original denominator in the answer.

Step 4: Change the improper fraction into a mixed numeral, and reduce to lowest terms.

$$4 \times \frac{2}{3} = \frac{8}{3} = 2\frac{2}{3}$$

In a rodeo, a bull rider must stay on the bull for at least $\frac{2}{15}$ minute.

No matter how big an iceberg, $\frac{8}{9}$ of it will hide below the surface of the water.

How to Multiply Mixed Numbers

Step 1: Change all mixed numerals to improper fractions.

Step 2: Multiply the numerator; the product is the new numerator.

Step 3: Multiply the denominator; the product is the new denominator.

Step 4: Change the improper fraction into a mixed numeral, and reduce to lowest terms.

$$7\frac{1}{2} \times 5\frac{2}{5} = \frac{15}{2} \times \frac{27}{5} = \frac{405}{10} = 40\frac{5}{10} = 40\frac{1}{2}$$

The *Titanic* was $882\frac{3}{4}$ feet long.

Dividing Fractions

How to Divide Fractions

Step 1: Invert the second fraction (the divisor fraction).

Front side

Step 2: Change the problem into a multiplication problem.

Flip side

Step 3: Multiply the fractions

Step 4: Reduce the quotient fraction to lowest terms.

$$\frac{8}{12} \div \frac{3}{4} = \frac{8}{12} \times \frac{4}{3} = \frac{32}{36} = \frac{8}{9}$$

How to Divide a Whole Number by a Fraction
(or a Fraction by a Whole Number)

Step 1: Change the whole number into an improper fraction with the whole number as the numerator and 1 as the denominator.

Step 2: Proceed with the instructions for dividing fractions.

Step 3: Change any improper fractions in the quotient to mixed numerals, and reduce to lowest terms.

$$5 \div \frac{7}{8} = \frac{5}{1} \div \frac{7}{8} = \frac{5}{1} \times \frac{8}{7} = \frac{40}{7} = 5\frac{5}{7}$$

How to Divide Mixed Numbers

Step 1: Change any mixed numbers into improper fractions.

Step 2: Proceed with the instructions for dividing fractions.

Step 3: Change any improper fractions in the quotient to mixed numerals, and reduce to lowest terms.

$$3\frac{1}{3} \div 4\frac{4}{5} = \frac{10}{3} \div \frac{24}{4} = \frac{10}{3} \times \frac{4}{24} = \frac{40}{72} = \frac{5}{9}$$

Ratio

A **ratio** is a comparison between two numbers or amounts. Ratios are used to compare all kinds of things, such as: age, prices, weights, times, or distances.

Help! There are 12 daddy longlegs, 5 tarantulas, and 6 rats in my basement!

Get Sharp Tip # 15

No matter how you write your ratio, it is always read the same way: *twelve to six*

What is the ratio of daddy longlegs to rats ?

You can write the ratio in three ways:

…as a fraction: 12/6   …with a colon: 12 : 6   …with words: 12 to 6

Terms of a Ratio

The numbers in a ratio are called **terms**.

In the ratios above, the terms are **12** and **6**.

The **12** is the first term and the **6** is the second term.

Watch the order of the terms carefully.

The above ratio 12 : 6 is the ratio of daddy longlegs to rats.

What is the ratio of rats to daddy longlegs?

The terms get reversed!
The ratio is 6 : 12 or 6/12.

*(Pay close attention to the **order** of terms! If you change the order, the meaning of the ratio is entirely different.)*

But what is the ratio of rats to tarantulas?

$\frac{6}{5}$

But what is the ratio of tarantulas to daddy longlegs?

$\frac{5}{12}$

What is the ratio of rats to the total number of creatures in the basement?

$\frac{6}{24}$

Better Grades & Higher Test Scores / MATH
©Incentive Publications, Inc., Nashville, TN

Reducing Ratios to Lowest Terms

Ratios are reduced in the same way that fractions are reduced.

The ratio of **rats to daddy longlegs** is 6/12. This can be reduced to ½.

The ratio of **rats to total creatures** is 6/24. This can be reduced to ¼.

Changing Fractions to Whole Number Ratios

You can change a fraction ratio to a whole number

The ratio of daddy longlegs to rats is 12/6, reduced to 2/1.

So you can say, *there are 2 daddy longlegs for every rat*, or *2 to 1*.

Equivalent Ratios

Equivalent ratios are ratios that name the same value.
When reduced to lowest terms, equivalent ratios will be the same.

7:56 is equivalent to **4:32**

7/56 in lowest terms is **1/8. 4/32** in lowest terms is **1/8.**

Using Equivalent Ratios to Find Missing Numbers

When you know the ratio, you can use equivalent ratios to find missing quantities.

In my neighborhood, there are 80 rat traps set in the 16 houses. Every house has the same number of traps. How many traps are in 1 house?

Step 1:
Write the ratio.

$$\frac{80}{16}$$

Step 2:
Find a ratio equivalent to $\frac{80}{16}$
Since the question asks how many traps in 1 house, the second term of the second ratio must be 1. To get 1, you must divide the first denominator (16) by 16.

Step 3:
Divide the 80 by 16 also. Write this as the first term in the new ratio.

$$\frac{80 \div 16}{16 \div 16} = \frac{5}{1}$$

5 traps in 1 house!

Rates

A **rate** is a ratio that compares quantities of different units.

Brie
$50 per
3 pounds

$$\frac{\$50}{3\ pounds}$$

80 MILES PER HOUR

$$\frac{80\ miles}{1\ hour}$$

5 FREE USED BOOKS PER $20 PURCHASE

$$\frac{5\ books}{\$20}$$

> **Get Sharp Tip # 16**
>
> The word you use to compare two units in a ratio is *per*.

Reducing Rates to Find Quantities

Charlie paid $11.20 for a 4-pound bag of vitamin-enriched rat food. What was the cost per pound?

Since you know the ratio (rate), you can find the cost by reducing the ratio to lowest terms. Divide both terms of the ratio by 4.
(You use 4 because the second term of the new ratio must be 1, so you must use the number that will yield a quotient of 1 when it goes into 4.)

$$\frac{\$11.20 \div 4}{4 \div 4} = \frac{\$2.80}{1}$$

It costs $ 2.80 per pound!

Using Equivalent Ratios to Find Other Rates

Since you know the ratio (rate), you can find other costs by writing equivalent ratios.

> How much does Charlie's rat food cost for 20 pounds?

Multiply both terms of the ratio by 5.
(You use 5 because the second term of the new ratio must be 20, so you must choose the number that will yield a product of 20 when multiplied by 4.)

$$\frac{\$11.20 \times 5}{4 \times 5} = \frac{\$56.00}{20}$$ *It costs $ 56.00 per 20 pounds!*

Better Grades & Higher Test Scores / MATH
©Incentive Publications, Inc., Nashville, TN

Proportions

A **proportion** is an equation which states that two ratios are equal.

I found 12 tarantulas by searching under 7 rocks. I figure that if I continue at the same rate, I'll need to look under 35 rocks to see 60 tarantulas.

This is the proportion for this idea:

$$\frac{12}{7} = \frac{60}{35}$$

Hey!

Solving Proportions

I also found 28 worms by searching under 4 rocks. At this rate, how many worms will I find under 60 rocks?

Proportions are very helpful in solving problems. They help to find missing numbers or quantities.

To find a missing number by solving a proportion, follow these steps:

Step 1:

Write the proportion. Use x for the missing number.

$$\frac{28}{4} = \frac{x}{60}$$

Step 2:

Cross multiply.

$28 \times 60 = 4x$

$1{,}680 = 4x$

Step 3:

Solve the equation for x.

$1{,}680 = 4x$

$420 = x$

Wow! 420 worms! That's almost enough to start a worm farm!

Decimals

A **decimal** is a way of writing a fractional number which has a denominator of 10 or a multiple of 10.
Decimals are written using a decimal point.
The decimal point is placed to the right of the ones' place.

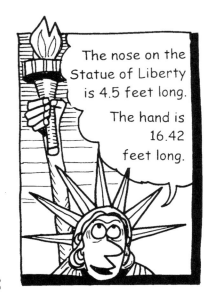

The nose on the Statue of Liberty is 4.5 feet long.

The hand is 16.42 feet long.

$$\frac{1}{10} = .1 \quad \frac{1}{100} = .01 \quad \frac{1}{1000} = .001 \quad \frac{1}{10,000} = .0001$$

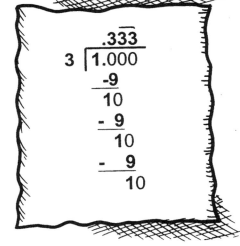

```
     .625
8 | 5.000
  - 48
    20
   -16
    40
   - 40
     0
```

Terminating Decimals

A **terminating decimal** is a decimal number that ends. When a quotient for a divided fraction eventually shows a remainder of zero, the decimal terminates.

When **5/8** is divided, the result is a terminating decimal.

```
     .3̄3̄3̄
3 | 1.000
   -9
   10
  - 9
   10
  - 9
   10
```

Repeating Decimals

A **repeating decimal** that has one or more digits which repeat indefinitely. The quotient for a divided fraction never results in a remainder of zero, and one or more of the final digits keep repeating. A repeating decimal is indicated by a bar written above the numbers that repeat.

When **1/3** is divided, the result is a repeating decimal.

Mixed Decimal Numbers

Mixed decimal numbers combine whole numbers and decimals.

A mixed number has digits on both sides of the decimal point.

Between 1990 and 1994, the Mississippi River received 702.5 million pounds of toxic discharge.

Phew!

Better Grades & Higher Test Scores / MATH
©Incentive Publications, Inc., Nashville, TN

Place Value in Decimals

Places to the right of the ones' place show decimals.
A decimal point separates the ones place from the tenths place.
The chart below shows the first six places to the right of the decimal point.

tens	ones	tenths	hundredths	thousandths	ten thousandths	hundred thousandths	millionths
	5.	5					
	1.	1	2	3			
	0.	0	0	7	1		
	0.	1	5	0	5	5	
1	2.	0	0	0	8	6	6

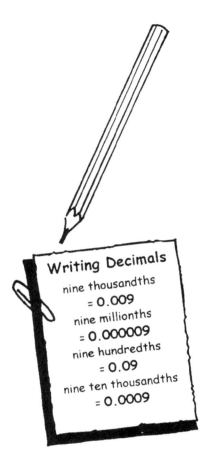

Writing Decimals

nine thousandths
= 0.009
nine millionths
= 0.000009
nine hundredths
= 0.09
nine ten thousandths
= 0.0009

Reading & Writing Decimals

Read the whole number first. Then read the entire number to the right
of the decimal point, adding the label from the place of the last digit.

5.5	reads	*five and five tenths*
1.123	reads	*one and one hundred twenty-three thousandths*
0.0071	reads	*seventy-one ten thousandths*
0.15055	reads	*fifteen thousand fifty-five hundred thousandths*
12.000866	reads	*twelve and eight hundred sixty-six millionths*

Rounding Decimals

Decimals are rounded in the same way as whole numbers.
If a digit is 5 or greater, round up to the next highest value in the place
to the left. If the digit is 4 or less, round down.

0.005 *rounds to* **0.01**

0.63 *rounds to* **0.6**

5.068 *rounded to the nearest tenth is* **5.1**

5.068 *rounded to the nearest hundredth is* **5.07**

Get Sharp: Decimals

Operations with Decimals

Adding & Subtracting Decimals

Step 1: Line up the decimal point in both numbers in the problem.

Step 2: Add or subtract just as with whole numbers.

Step 3: Align the decimal point in the sum or difference with decimal points in the numbers above.

International Falls, Minnesota is the coolest U.S. town—with average temperatures of 36.4° F. Key West, Florida has the warmest average. That temperature is 77.7° F. What's the difference?

$$\begin{array}{r} 77.7° \\ -\ 36.4° \\ \hline 41.3° \end{array}$$

The neighbors are complaining about the 97.8° F temperatures today. But the temperature on the sun's surface is 9,529° hotter. What is the sun's temperature?

$$\begin{array}{r} 97.8° \\ +\ 9,529.0° \\ \hline 9,626.8° \end{array}$$

The driest city in the U.S. is Yuma, Arizona, with 2.65 inches of precipitation annually.
The wettest city, Quillayute Washington has about 39.6 times as much. About how much moisture falls in the Washington city?

$$\begin{array}{r} 2.65 \\ \times\ 39.6 \\ \hline 1590 \\ 2385 \\ +\ 795 \\ \hline 104.940 \end{array}$$

Multiplying Decimals

Step 1: Multiply as you would with whole numbers.

Multiply 2.65 x 3.96 to get 104940.

Step 2: Count the number of places to the right of the decimal point in both factors (total).

Count the number of places to the right of the decimal point: 2.65 has 2; 39.6 has 1, for a total of 3.

Step 3: Count over from the right end of the product that same number of places.

In the product, count 3 places backward from the right.

Step 4: Insert the decimal point.

Place the decimal point 4 and the 9.
Quillayute's annual precipitation is about 104.940 inches.

Dividing a Decimal by a Whole Number

Step 1: Place the decimal point in the quotient directly above the decimal point in the dividend.

Step 2: Divide as you would with whole numbers.

Step 3: Add zeros where necessary to hold places.

Daring Dan climbed an frozen waterfall that was 236.7 feet tall. He climbed over a period of 3 days. What was the average distance he climbed each day?

$$
\begin{array}{r}
78.9 \text{ ft.} \\
3\overline{)236.7} \\
-21 \\
\hline
26 \\
-24 \\
\hline
27 \\
-27 \\
\hline
0
\end{array}
$$

Falling water! Look out below!

Angel Falls in Venezuela is the world's tallest falls. The ocean's deepest trench is the Marianas Trench at 35,820 feet. This trench is 11.15 times the height of the falls. How tall is Angel Falls?
(Round to the nearest whole number.)

$$
\begin{array}{r}
3212.5 \\
11.15\overline{)35,820.00} \\
-3345 \\
\hline
2370 \\
-2230 \\
\hline
1400 \\
-1115 \\
\hline
2850 \\
-2230 \\
\hline
6200 \\
-5575 \\
\hline
625
\end{array}
$$

answer: 3213 feet

Dividing by a Decimal

Step 1: Move the decimal point to the right to write the divisor as a whole number. Count the number of places you must move the decimal point.

Step 2: Move the decimal point in the dividend the same number of places to the right.

Step 3: Divide as you would with whole numbers.

Step 4: Align the decimal point in the quotient with the decimal point in the dividend.

Better Grades & Higher Test Scores / MATH
©Incentive Publications, Inc., Nashville, TN

Get Sharp: Decimals

Money

Decimals are used to write amounts of money.
Money is shown in decimal numbers to the hundredth place.
In a money amount, the decimal point is placed after the dollars
(to the right of the one dollars' place).

The first place to the right of the decimal is the place of ten cents
(1-tenth of a dollar).

The second place to the right of the decimal is the place of one cents
(1-hundredth of a dollar).

Get Sharp Tip # 17

When you do calculations with money, always round off amounts to the nearest hundred (cents).

$$\$\ 142.56$$

ten cents place (one 10th of a dollar)
one cents' place (one 100th of a dollar)

$99.09   reads *ninety-nine dollars and nine cents*

$9.83   reads *nine dollars and eighty-three cents*

$27.15   reads *twenty-seven dollars and fifteen cents*

$100.30   reads *one hundred dollars and thirty cents*

$ 0.47   reads *forty-seven cents*

There are many combinations of coins and bills that make up amounts of money:

2 ten-dollar bills

+ 1 five-dollar bill

+ 7 quarters

$16.75

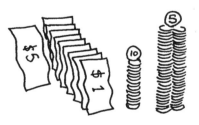

1 five-dollar bill

+ 8 one-dollar bills

+ 10 dimes

+ 55 nickels

$16.75

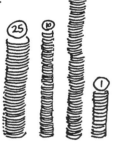

36 quarters

+ 44 dimes

+ 65 nickels

+ 10 pennies

$16.75

Operations with Money

Operations with money are just like operations with decimals, because money amounts are decimals.

Addition

Line up the decimal points carefully in both addends. Align the decimal point in the sum with these.

> On its opening weekend, the movie, *Spiderman*, took in a record $114,800,000.00. That same weekend, *Ice Age* took in $2,400,000.00. How much did they make together?

$$\$114,800,000.00$$
$$+ \quad 2,400,000.00$$
$$\overline{\$117,200,000.00} \ !$$

Subtraction

Align the decimal points in both numbers. Align the decimal point in the difference with these.

> *Harry Potter and the Sorcerer's Stone* took in $90,300,000.00 on its first weekend. How much less was that total than *Spiderman's* opening weekend?

$$\$\ 114,800,000.00$$
$$- \quad 90,300,000.00$$
$$\overline{\$\ 24,500,000.00}$$

Multiplication

Multiply as with whole numbers. Tally the total number of places to the right of the decimal point. Count the same number of places from the right in the product.

> Charlie took 12 friends to see *Spiderman*. He spent $13.50 on each friend (for a ticket, popcorn, and drink.) What was the total amount generous Charlie spent?

$$\$\ 13.50$$
$$\times 12$$
$$\overline{2700}$$
$$+ 1350$$
$$\overline{\$162.00}$$

Division

Move the decimal point in the divisor to make it a whole number. Move the decimal point in the dividend the same number of places. Align the decimal point in the quotient with the decimal point in the dividend. Divide as with whole numbers.

> Last year, Abby bought Harry Potter souvenirs every month. She spent $219.60 during the year. On the average, what did she spend each month?

$$
\begin{array}{r}
\$18.30 \\
12\overline{)\$219.60} \\
-12 \\
\hline
99 \\
-96 \\
\hline
36 \\
-36 \\
\hline
00
\end{array}
$$

Better Grades & Higher Test Scores / MATH
©Incentive Publications, Inc., Nashville, TN

Get Sharp: Money

Fractions & Decimals

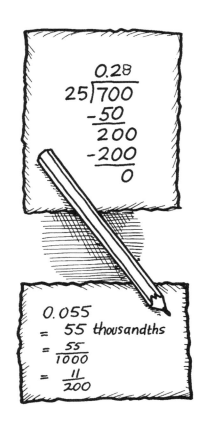

$$0.28$$
$$25 \overline{)700}$$
$$\underline{-50}$$
$$200$$
$$\underline{-200}$$
$$0$$

$$0.055$$
$$= 55 \text{ thousandths}$$
$$= \frac{55}{1000}$$
$$= \frac{11}{200}$$

How to Write a Fraction as a Decimal

Step 1: Divide the numerator by the denominator.

Step 2: Write a zero to hold the ones' place
(if there is no number in that place).

$$7/25 = 0.28$$

How to Write a Decimal as a Fraction

Step 1: Remove the decimal point and write the number as the numerator. The denominator is 10 or a multiple of 10, depending what place the last digit of the decimal occupied. For instance, in 0.355, the last digit is a thousandth.

Step 2: Reduce the fraction to lowest terms.

WHO WON?

Alexi ran the obstacle course in 7.38 minutes.

Yanni ran the course in 7 ¾ minutes.

Ruben ran the course in 0.15 hours.

Change everything to decimals in terms of minutes. Then compare the times.
7.38 min = 7.38 min
7 ¾ min = 7.75 min
0.15 hours = 0.15 × 60 = 9 min.

Alexi's time is fastest!

Better Grades & Higher Test Scores / MATH
©Incentive Publications, Inc., Nashville, TN

Percent

Percent is a ratio written as a decimal. In a percent the ratio always compares a number to 100. Percent means *per one hundred* or *one part in a hundred*. Percent is shown with the symbol %. Any fraction written with a denominator of 100 can be changed into a percent.

Get Sharp Tip # 18

Remember this: Decimal to Percent— move right! Percent to Decimal— move left!

How to Write a Decimal as a Percent

Move the decimal point **two places to the right**.

$$0.465 = 46.5\% \qquad 15.4 = 1540\%$$

How to Write a Percent as a Decimal

Move the decimal point **two places to the left**.

$$19.6\% = 0.196 \qquad 124\% = 12.4$$

How to Write a Fraction as a Percent

Any fraction with a denominator of 100 can easily be changed to a percent.

Drop the denominator and write the % symbol.

$$\frac{47}{100} = 47\%, \qquad \frac{8}{100} = 8\%$$

For other fractions:

Divide the numerator by the denominator. Then move the decimal point two places to the right.

$$\frac{3}{10} = 3 \div 10 = 0.30 = 30\%$$
$$\frac{4}{5} = 4 \div 5 = 0.80 = 80\%$$

(The decimal point moved two places to the right.)

60 out of every 100 members of the American population watches movies for entertainment on a regular basis.

99% of all American households have at least one television set.

Is that 6%? 60%? 06%? 600%?

It's 60%

It's 0.99

Is that 0.99? 99.00? 9.9? 0.9?

Better Grades & Higher Test Scores / MATH
©Incentive Publications, Inc., Nashville, TN

Get Sharp: Percent

How to Find a Percentage of a Number

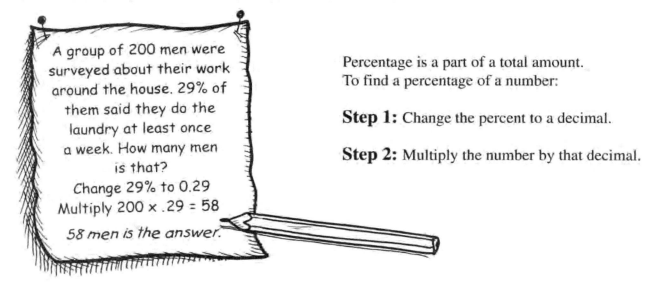

A group of 200 men were surveyed about their work around the house. 29% of them said they do the laundry at least once a week. How many men is that?
Change 29% to 0.29
Multiply 200 x .29 = 58

58 men is the answer.

Percentage is a part of a total amount.
To find a percentage of a number:

Step 1: Change the percent to a decimal.

Step 2: Multiply the number by that decimal.

How to Tell What Percent One Number Is of Another

To tell what percent one number is of a second number:

Step 1: Divide the first number by the second.

Step 2: Express your answer as a percent.

In a group of 450 men, 90 said they had proposed marriage on their knees.
What percent of the men is that?

Divide 90 by 450 = 0.20

The answer is 20%.

How to Find the Base When You Know the Percent

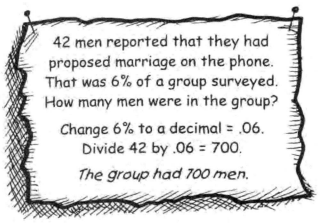

42 men reported that they had proposed marriage on the phone. That was 6% of a group surveyed. How many men were in the group?

Change 6% to a decimal = .06.
Divide 42 by .06 = 700.

The group had 700 men.

When you know the percent and the resulting percentage (part of the total number), but not the base number (the original number):

Step 1: Change the percent to a decimal.

Step 2: Divide percentage number by that decimal.

Taxes, Tips, & Discounts

There are many times when you need to find a percentage of a number—to calculate taxes on a purchase, to find savings at a sale, to offer discounts on services, to figure a tip for your waiter.

To figure percentage of an amount, always:

Step 1: Change the percent rate to a decimal.

Step 2: Multiply the original amount by the decimal.

COOL-TO-THE-MAX MALL

SAXOPHONE SALE $325 original price 25% OFF

Discount on all DRUM SETS 30%

The sales tax in my cousin's state is 5%. Guess how much tax he paid on his new guitar!
$230 x .05 = $11.50
Not bad!

Zeke's dad bought him a $500 drum set on sale. How much did he pay after the discount?
$500 x .30 = $150
$500 - $150 = $350
He got the drums for $350!

After lunch, I want to buy a saxophone on sale. How much will I save?
$325 x .25 = $81.25

Receipt
soup   3.40
salad   2.75
muffin   1.20
milkshake   2.65
            $ 10.00
Thank you!

My lunch bill is $10! I should leave a 15% tip. How much will that be?
$10 x .15 = $1.50
Hey, can I borrow $11.50?

Flip's Flops and Footwear

SO CHIC

Simple Interest

When you're earning interest, you'll be interested in having interest rates go up.

The **principal** is the amount of money that you deposit in a bank or that you loan or borrow.

The **interest** is an amount of money paid or earned on a bank deposit or a loan.

The **rate of interest** is the percent of the principal earned or charged over a certain period of time (usually a year).

Time is the period over which the interest is figured.

When you're paying interest, you'll be interested in having interest rates go down.

How to Find Simple Interest

Step 1: Change the interest percent to a decimal.

Step 2: Use this formula: I *(Interest)* = P *(Principal)* x R *(Rate of Interest)* x T *(Time)*

Step 3: Insert the numbers (including the decimal for R) and multiply.

Rufus deposited $800 in a new bank account. The bank will give him 1.5% per year. How much simple interest will Rufus earn in 3 years?

GOOD KARMA BANK

$800.
x .015
4000
800
$12.000 x 3
= $36.00

Carly bought a new keyboard It cost $750. She did not start paying until after 2 years. The interest rate was 6%. How much interest will collect in 2 years?

$750. x .06 x 2 = $90.00

If Carly pays the principal and the interest after 2 years, how much will the keyboard have cost her?

$750.
+ 90.
$840.

GET SHARP→

on

Geometry & Measurement

Points, Lines, & Planes

What's the point?

A **point** is a precise location in space.

It has no thickness, but a dot can be used to represent a point so that it can be seen.

A point is named with a letter.

It is represented like this: point A, point B.

•A   B•

A **line** is a set of points that extends indefinitely in two opposite directions.

A line is named by any two points along it.

A line is represented like this: $\overleftrightarrow{CD}$.

Intersecting lines cross each other.

Lines EF and GH intersect at point I.

$\overleftrightarrow{EF}$ and $\overleftrightarrow{GH}$ intersect at point **I**.

Parallel lines never intersect because they are always the same distance apart.

$\overleftrightarrow{JK}$ is parallel to $\overleftrightarrow{LM}$.

Perpendicular lines intersect at right angles.

$\overleftrightarrow{OP}$ is perpendicular to $\overleftrightarrow{QR}$.

Four right angles are formed when they intersect.

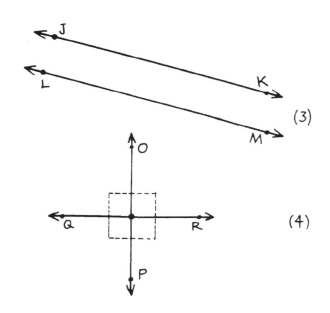

A **line segment** is a part of a line.

It has two endpoints.

A line segment is named by its endpoints.

A line segment is represented like this: $\overline{ST}$

X is the **midpoint** of $\overline{ST}$.

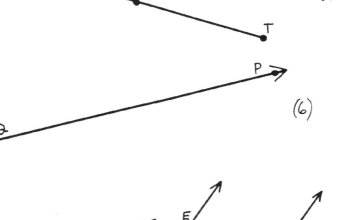

(5)

(6)

A **ray** is a part of a line that has one endpoint and expends indefinitely in one direction.

A ray is named by the endpoint and another point along the line.

A ray is represented like this: $\overrightarrow{QP}$

A **plane** is a set of all the points on a flat surface that extends indefinitely in all directions.

A plane is named by four letters.

A plane is represented like this: **plane EFGH**.

Intersecting planes cross each other, meeting along a line.

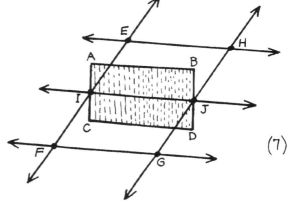

(7)

Plane ABCD intersects **plane EFGJ** along $\overleftrightarrow{IJ}$.

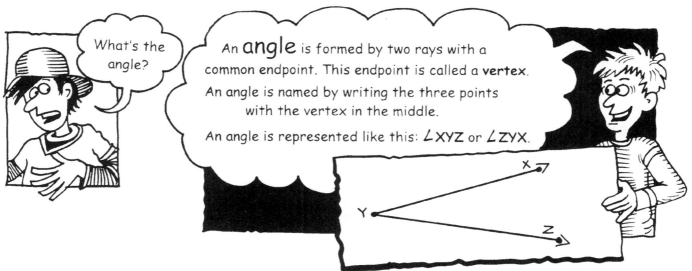

What's the angle?

An **angle** is formed by two rays with a common endpoint. This endpoint is called a **vertex**.

An angle is named by writing the three points with the vertex in the middle.

An angle is represented like this: ∠XYZ or ∠ZYX.

Get Sharp: Plane Geometry

Angles

An **angle** is a figure formed by two rays with a connecting **endpoint**.

This endpoint is called a **vertex**.

Angles come in many sizes.

Angles are classified by their measurements.

The unit of measurement for an angle is a **degree**.

A **protractor** is used to measure angles.

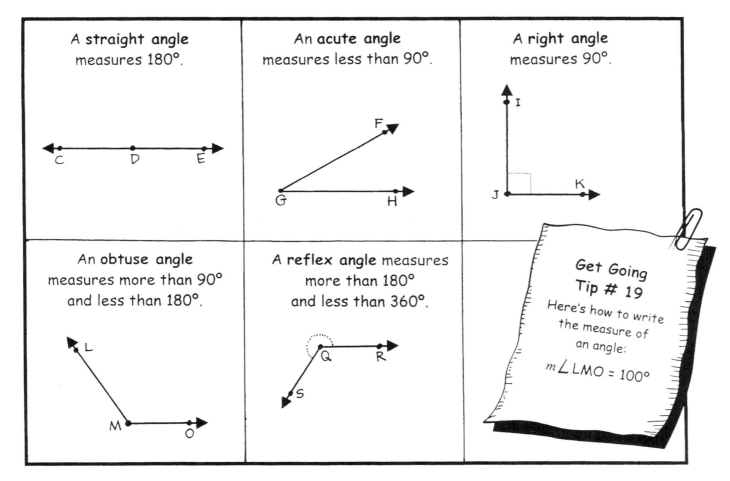

A **straight angle** measures 180°.

An **acute angle** measures less than 90°.

A **right angle** measures 90°.

An **obtuse angle** measures more than 90° and less than 180°.

A **reflex angle** measures more than 180° and less than 360°.

Get Going Tip # 19

Here's how to write the measure of an angle:

$m\angle LMO = 100°$

Angle Relationships

Angles have some very curious and interesting relationships with one another.

Make sure you can keep all these relationships straight!

Congruent Angles (A)

Angles that have the same measure are congruent.
∠ CDF and ∠ FDE are congruent because they both measure 45°.
Write ∠CDF ≅ ∠FDE.

Complementary Angles (B)

When the sum of two angles is 90°, the angles are complementary.
∠ GHJ and ∠ JHI are complementary.

Supplementary Angles (C)

When the sum of two angles is 180°, they are supplementary.
∠KLN and ∠NLM are supplementary.

Adjacent Angles (D)

When angles have a common vertex and a common edge (or leg), they are adjacent.
∠OPR and ∠RPQ are adjacent.

Vertical Angles (E)

When two lines intersect, vertical angles are formed. These are the angles that are opposite each other at the vertex.

Two pairs are formed by the intersection of $\overline{ST}$ and $\overline{UV}$.
These pairs are: ∠SWU & ∠VWT; ∠SWV and ∠UWT.

Get Sharp Tip # 20
Vertical angles are congruent (have the same measure).

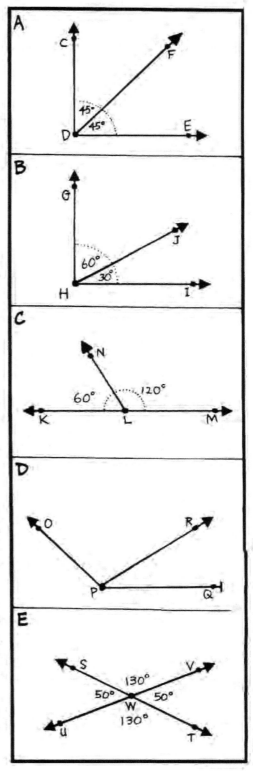

Parallel Lines & Angles

A third line will intersect sometimes two parallel lines. This third line is called a **transversal**
Several pairs of angles are formed around the intersecting lines. Many of the pairs are congruent.

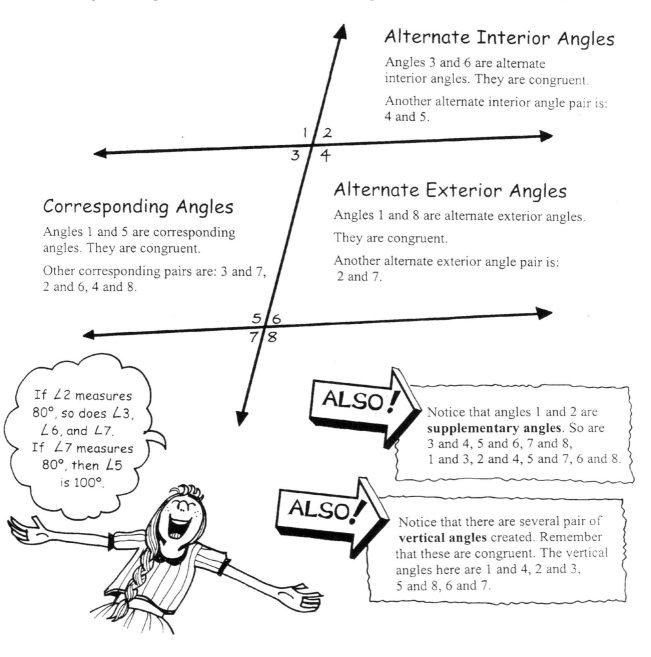

Alternate Interior Angles

Angles 3 and 6 are alternate interior angles. They are congruent.

Another alternate interior angle pair is: 4 and 5.

Alternate Exterior Angles

Angles 1 and 8 are alternate exterior angles.

They are congruent.

Another alternate exterior angle pair is: 2 and 7.

Corresponding Angles

Angles 1 and 5 are corresponding angles. They are congruent.

Other corresponding pairs are: 3 and 7, 2 and 6, 4 and 8.

If ∠2 measures 80°, so does ∠3, ∠6, and ∠7.
If ∠7 measures 80°, then ∠5 is 100°.

ALSO! Notice that angles 1 and 2 are **supplementary angles**. So are 3 and 4, 5 and 6, 7 and 8, 1 and 3, 2 and 4, 5 and 7, 6 and 8.

ALSO! Notice that there are several pair of **vertical angles** created. Remember that these are congruent. The vertical angles here are 1 and 4, 2 and 3, 5 and 8, 6 and 7.

Better Grades & Higher Test Scores / MATH
©Incentive Publications, Inc., Nashville, TN

Bisecting Line Segments & Angles

When a line segment or angle is **bisected**,
it is divided into two congruent parts (parts of equal length).
The **bisector** is the line that divides the line segment
or the ray that divides the angle.

1. Open a compass to a radius greater than half the line.

2. Use **A** as the center. Draw an arc intersecting $\overline{AB}$.

3. Use **B** as a center. Draw an arc intersecting $\overline{AB}$.

4. Label the points where the two arcs intersect **X** and **Y**.

5. Draw line $\overleftrightarrow{XY}$ to connect the two points **X** and **Y**.

$\overleftrightarrow{XY}$ is a perpendicular bisector of $\overline{AB}$.

1. Using the vertex **Y** as the center, use a compass to draw an arc intersecting $\overrightarrow{YX}$ and $\overrightarrow{YZ}$.

2. Use **S** as a center. Open the compass to a radius greater than half the distance between **S** and **T**.

3. Use **S** as a center. Draw an arc in the space between $\overrightarrow{YX}$ and $\overrightarrow{YZ}$.

4. Keep the same compass opening. Use **T** as a center and repeat step 3.

5. Label the point where the arcs intersect as point **V**.

6. Draw a new ray $\overrightarrow{YV}$.

$\overrightarrow{YV}$ is the bisector of **XYZ** (or **ZYX**).

Polygons

A **polygon** is a plane figure formed by joining three or more straight line segments at their endpoints. The line segments are called **sides**. Each endpoint where sides join is a **vertex**.

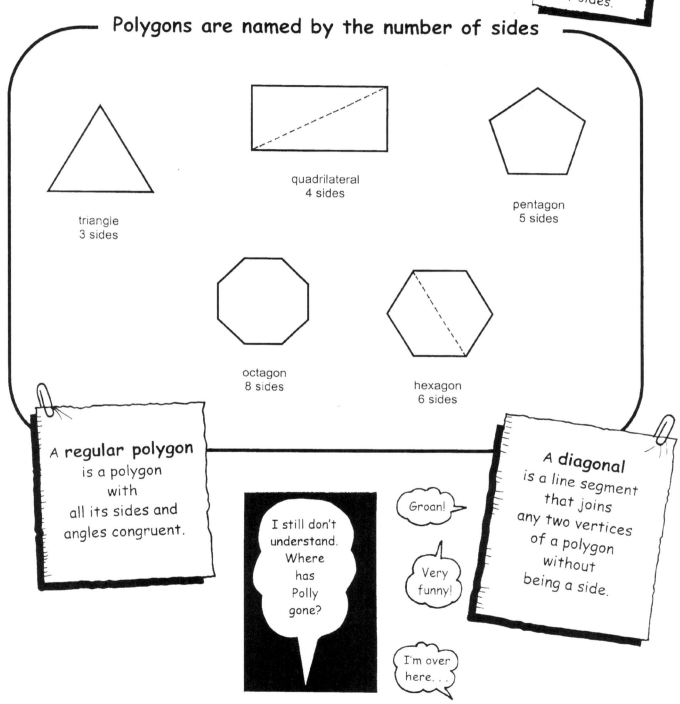

Polygons are named by the number of sides

triangle
3 sides

quadrilateral
4 sides

pentagon
5 sides

octagon
8 sides

hexagon
6 sides

A **regular polygon** is a polygon with all its sides and angles congruent.

I still don't understand. Where has Polly gone?

Groan!

Very funny!

I'm over here. . .

A **diagonal** is a line segment that joins any two vertices of a polygon without being a side.

Triangles

A **triangle** is a three-sided plane figure. But not all triangles are the same!

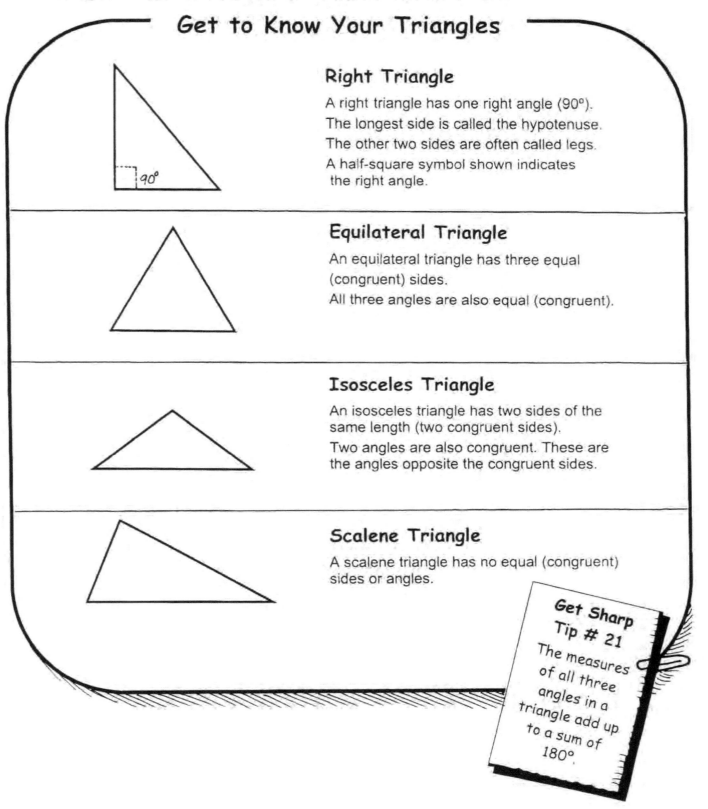

Get to Know Your Triangles

Right Triangle

A right triangle has one right angle (90°).
The longest side is called the hypotenuse.
The other two sides are often called legs.
A half-square symbol shown indicates
the right angle.

90°

Equilateral Triangle

An equilateral triangle has three equal
(congruent) sides.
All three angles are also equal (congruent).

Isosceles Triangle

An isosceles triangle has two sides of the
same length (two congruent sides).
Two angles are also congruent. These are
the angles opposite the congruent sides.

Scalene Triangle

A scalene triangle has no equal (congruent)
sides or angles.

Get Sharp Tip # 21
The measures of all three angles in a triangle add up to a sum of 180°.

Better Grades & Higher Test Scores / MATH
©Incentive Publications, Inc., Nashville, TN

Get Sharp: Plane Geometry

Quadrilaterals

A **quadrilateral** is a polygon that has four sides and four vertices.
Don't be confused by the different kinds of quadrilaterals.

A parallelogram is a quadrilateral that has parallel line segments in both pairs of opposite sides.

Opposite internal angles are congruent.

Opposite sides are congruent.

A rectangle is a parallelogram that has four right angles.

A square is a rectangle that has sides of equal length and four right angles.

A square is a parallelogram.

A rhombus is a parallelogram that has four sides of equal length.

A rhombus does not necessarily have four right angles.

A trapezoid is a quadrilateral that has only one pair of parallel sides.

An isosceles trapezoid is a trapezoid that has congruent nonparallel sides and congruent base angles.

Circles

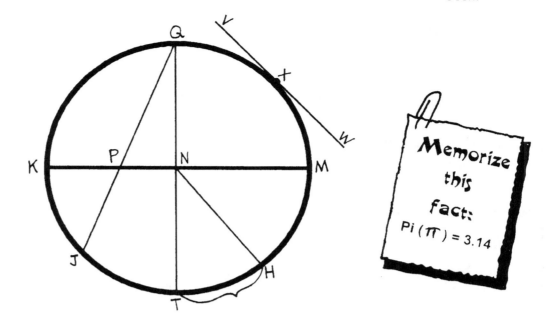

A circle is the set of all the points in a plane that are the same distance from a particular point. That given point is the center of the circle *(N)*.

An arc is any part of the circle or any section of the line segment that forms the outside edge of the circle.

(Example: H to T is an arc.)

The radius is any line segment from the center to a point on the circumference of the circle.

(Examples: $\overline{MN}$, $\overline{NT}$, $\overline{NK}$, $\overline{NQ}$, $\overline{NH}$ are radii.)

A central angle is an angle formed by two radii of a circle.

(Example: ∠ TNK and ∠ HNM are central angles.

The diameter is a chord through the center of the circle.
The diameter is twice the length of a radius.
(Examples: $\overline{QT}$ and $\overline{MK}$ are diameters.)

A chord is a line segment joining two points on the circle.
(Example: $\overline{QJ}$ and $\overline{QT}$ are chords.)

A tangent is a line that touches the edge of a circle at one point but does not pass into or through the circle.
(Example: $\overleftrightarrow{VW}$ touches the circle at point X)

The circumference is the distance around the outside edge of the circle.
The circumference of a circle is the length of its diameter multiplied by **π** *(pi, which is approximately equal to 3.14).*

The original Hula Hoop comes in three sizes: diameters of 30, 32, and 34 inches. In the first two months after the Wham-O Company introduced the hula hoop, 25 million were sold. 100 million orders from around the world followed quickly.

Better Grades & Higher Test Scores / MATH
©Incentive Publications, Inc., Nashville, TN

Get Sharp: Plane Geometry

Congruent & Similar Figures

Congruent Figures are figures that have the same size and shape. Each angle in one figure has a corresponding angle of the same measure. Each side in one figure has a corresponding side of the same length.

Similar Figures

Similar figures have congruent corresponding angles, but the corresponding sides are not necessarily congruent.

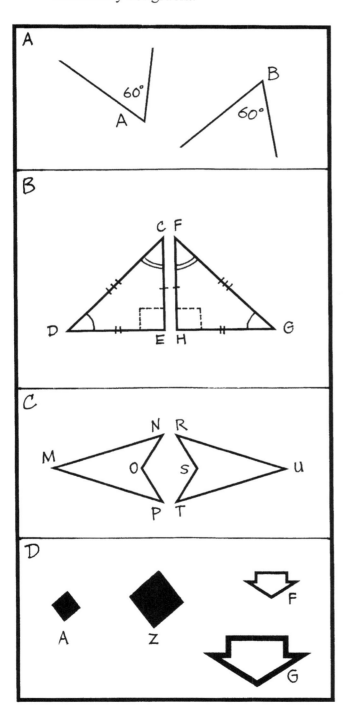

Congruent Angles (A)

Two angles are congruent when they have the same measure. Angle A is congruent to angle B because they have the same measure.

Write: $\angle A \cong \angle B$

Congruent Triangles (B)

These triangles are congruent—they have the same size and shape. The corresponding angles are congruent. (They have the same measure.) The corresponding sides are congruent. (They have the same length.) Notice how the angles and sides are marked alike to show the congruence.

Write: $\triangle CDE \cong \triangle FGH$

Other Congruent Polygons (C)

These two polygons are congruent. For each side and angle in figure MNOP, there is a corresponding congruent side or angle in figure URST

Write: Figure MNOP $\cong$ URST

Similar Figures (D)

The angles in figure A are congruent with corresponding angles in figure Z. The lengths of the sides, however, are different.

$\Diamond$A is similar to $\Diamond$Z
Figure F is similar to figure G.

Write: A $\sim$ Z
F $\sim$ G

Get Sharp: Plane Geometry

Better Grades & Higher Test Scores / MATH
©Incentive Publications, Inc., Nashville, TN

Tessellations, Transformations, & Symmetry

Tess Tells the Tessellation Tale

Tessellations are patterns formed by repeating a polygon. The same figure repeats in a pattern with no spaces and no overlaps. In a tessellation, the sum of the measures of the angles meeting at a common vertex is 360°.

Transformations

Transformations are changes in the position of a figure. There are different kinds of transformations.

A **slide** (or translation) is a move of a figure to another position.

A **turn** (or rotation) is a movement or pivot of a figure around a point.

A **flip** (or reflection) is a reversal of a figure along the line of one side.

Symmetry

When a figure can be divided into two congruent parts by a line, it is said to be **symmetrical**. (The line is called **the line of symmetry**.)

This is.

This is not.

Better Grades & Higher Test Scores / MATH
©Incentive Publications, Inc., Nashville, TN

Get Sharp: Plane Geometry

Space Figures

Space figures are geometric figures with three dimensions (length, width, and height). Most space figures have faces (sides that are polygons), vertices (points where more than two sides meet), and edges (lines along which two sides meet). Most also have a base. (The base is one of the faces.)

A **polyhedron** is a space figure with polygonal faces.

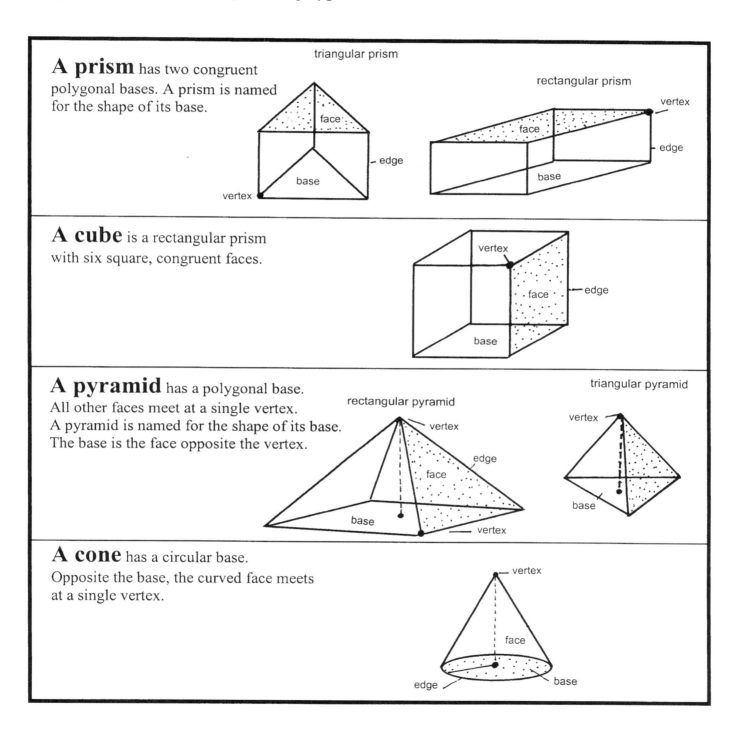

A prism has two congruent polygonal bases. A prism is named for the shape of its base.

A cube is a rectangular prism with six square, congruent faces.

A pyramid has a polygonal base. All other faces meet at a single vertex. A pyramid is named for the shape of its base. The base is the face opposite the vertex.

A cone has a circular base. Opposite the base, the curved face meets at a single vertex.

Better Grades & Higher Test Scores / MATH
©Incentive Publications, Inc., Nashville, TN

A cylinder has two congruent circular bases. A curved face joins the two bases.

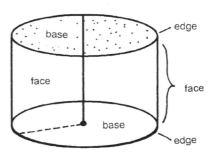

A sphere is a space figure with curved edges. Every point on the edge is an equal distance from the center in all four directions. A sphere has no base and no polygonal sides.

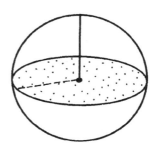

Euler (pronounced **Oiler**) was an extremely prolific mathematician. He developed theorems and tested ideas even into his older years. With this formula, Euler showed the relationship between the number of faces (F), vertices (V), and edges (E) in many polyhedrons.

Try out the formula with these polyhedons.

Figure	Shape of Base	Number of faces (F)	Number of Vertices (V)	Number of Edges (E)
cube	square	6	8	12
rectangular prism	rectangle	6	8	12
triangular prism	triangle	5	6	9
pentagonal prism	pentagon	7	10	15
octagonal prism	octagon	10	16	24
triangular pyramid	triangle	4	4	6
rectangular pyramid	rectangle	5	5	8
hexagonal pyramid	hexagon	7	7	12

Euler's Formula
$$V + F - E = 2$$

Measuring Weight

An object's **weight** (or mass) has to do with the amount of force (from Earth's gravity) pulling on it.

Scales are the tools used to measure weight. Weight is measured in different units. Some of the most common are ounces, pounds, tons, grams, kilograms, and metric tons.

U.S. Customary System Units of Weight
(basic unit – pound)

16 ounces (oz)	=	1 pound (lb)
2000 pounds (lbs)	=	1 Ton (T)

Metric System Units of Weight
(basic unit – gram)

1 gram (g)	=	1000 milligrams (mg)
1 gram (g)	=	100 centigrams (cg)
1 gram (g)	=	10 decigrams (dg)
1 kilogram (kg)	=	1000 grams (g)
1 metric ton (t)	=	1000 kilograms (kg)

Some Approximate Weights

1 g	a large feather
2.5 g	a ping pong ball
3 g	tiny banana bat
1.5 oz	a golf ball
.4 oz	a house mouse
.05 oz	a pygmy shrew
1.6 g	the smallest hummingbird
15 oz	a football
12 lb	a fat cat
30 kg	a medium-sized monkey
45 lbs	an average 6-year old child
62.4 lbs	1 cubic foot of water
485 lbs	a large gorilla
355 kg	a giant sea turtle
343 lbs	an ostrich *(world's heaviest bird)*
3500 kg	a great white shark
1.5 T	a Stegosaurus dinosaur
2.5 oz	the brain of a Stegosaurus
5.5 T	African elephant
150 T	the blue whale *(world's heaviest animal)*
90,854 T	the Washington Monument
5,216,400 t	the Great Pyramid
80 trillion T	planet Earth

HOW DO THEY COMPARE?

A pound is about 454 grams.
A kilogram is about 2.2 pounds.

On Earth, my 200-lb gorilla weighs about 90 kg.
On the moon, the pull of gravity is weaker. There, he would weigh about 34 lbs or 15 kg.

Better Grades & Higher Test Scores / MATH
©Incentive Publications, Inc., Nashville, TN

Measuring Length

Length is the distance along a line from one point to another.

Length is measured with many tools, such as rulers, meter sticks, yardsticks, and odometers.

Some Approximate Lengths

25.4 mm	thickness of hockey puck
6.5 cm	a house mouse
.5 in	the tiniest frog
15 in	stick insect *(longest insect)*
6.5 ft	a large gorilla
2.44 m	a giraffe
5 m	Nile crocodile
12 yd	reticulated python
15 m	humpback whale
57 ft	a giant deep sea squid
63 ft	wire in a standard *Slinky™*
882.75 ft	the ship, *Titanic*
1991 m	longest suspension bridge
3212 ft	world's highest waterfall
6670 km	world's longest river
40.3 mi	the Panama Canal
1052 ft	Eiffel Tower
5315 ft	world's deepest lake
9186 ft	depth of ice at South Pole
8848 m	world's highest mountain
3088 mi	Boston to San Diego
4,374,000 km	the sun's diameter
25 billion mi	Earth to the nearest star

U.S. Customary System Units of Length
(basic unit – foot)

1 foot (ft)	= 12 inches (in)
1 yard (yd)	= 3 ft or 36 in
1 rod (rd)	= 5½ yd
1 furlong (fur)	= 40 rd or 220 yd or 660 ft
1 mile (m)	= 5280 ft or 1760 yd

Metric System Units of Length
(basic unit – meter)

1 meter (m)	=	1000 millimeters (mm)
1 meter (m)	=	100 centigrams (cm)
1 meter (m)	=	10 decigrams (dm)
1 dekameter (dcm)	=	10 m
1 hectometer (hm)	=	100 m
1 kilometer (km)	=	1000 m

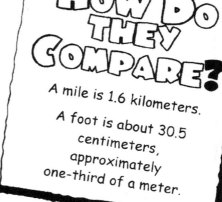

HOW DO THEY COMPARE?

A mile is 1.6 kilometers.

A foot is about 30.5 centimeters, approximately one-third of a meter.

I was just chased 10 miles by a 200-pound gorilla! That's 6.2 kilometers!

Measuring Perimeter & Circumference

Perimeter is the distance around any polygon.
To find the perimeter of a polygon, add the lengths of all sides,
or use a formula for a shortcut.

Circumference is the distance around a circle.
To find the circumference of a circle, use a formula.

Square	P = 4s	Other Polygons	P = sum of all sides
Rectangle	P = 2 l (length) + 2 w (width)	Circle	C = πd or π2r
Triangle	P = s + s + s		

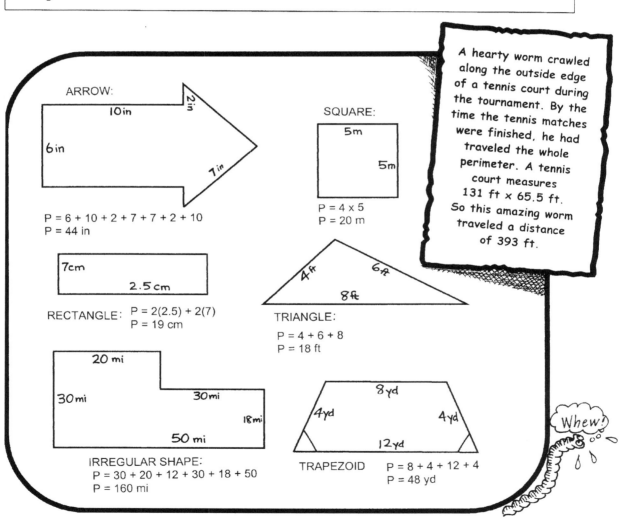

ARROW:

10 in
6 in
2 in
7 in

P = 6 + 10 + 2 + 7 + 7 + 2 + 10
P = 44 in

SQUARE:

5 m
5 m

P = 4 x 5
P = 20 m

7 cm
2.5 cm

RECTANGLE: P = 2(2.5) + 2(7)
P = 19 cm

TRIANGLE:
4 ft
6 ft
8 ft

P = 4 + 6 + 8
P = 18 ft

20 mi
30 mi
30 mi
18 mi
50 mi

IRREGULAR SHAPE:
P = 30 + 20 + 12 + 30 + 18 + 50
P = 160 mi

TRAPEZOID
8 yd
4 yd
4 yd
12 yd

P = 8 + 4 + 12 + 4
P = 48 yd

A hearty worm crawled along the outside edge of a tennis court during the tournament. By the time the tennis matches were finished, he had traveled the whole perimeter. A tennis court measures 131 ft x 65.5 ft. So this amazing worm traveled a distance of 393 ft.

Whew!

Circumference is the distance around the outside of a circle.

The measure of the circumference is the diameter (twice the radius) multiplied by pi ($\pi = 3.14$).

Here's the formula: $C = \pi d$ or $\pi 2r$

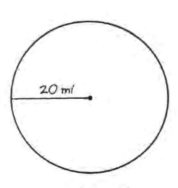

20 mi

$C = \pi(2 \times 20)$
$C = \pi(40)$
$C = 125.6$ mi.

DID YOU EVER WONDER?

IF YOU JOG AROUND THE OUTSIDE OF A WRESTLING MAT, HOW FAR HAVE YOU TRAVELED?
THE RADIUS IS 6.5m.
$3.14 \times 2(6.5) = 40.92m$

YOU'VE JOGGED 40.92 METERS!

Get Sharp: Measurement

Measuring Area of Polygons

Area is the size of the flat surface contained within a polygon or circle. Area is measured in square units.

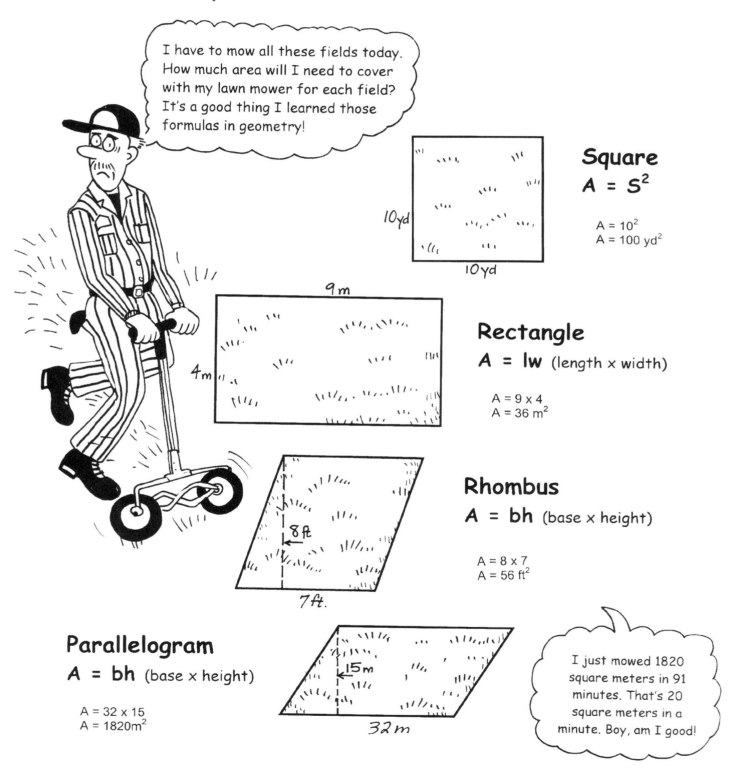

I have to mow all these fields today. How much area will I need to cover with my lawn mower for each field? It's a good thing I learned those formulas in geometry!

Square
$A = S^2$

10 yd

10 yd

$A = 10^2$
$A = 100 \text{ yd}^2$

9 m

4 m

Rectangle
$A = lw$ (length x width)

$A = 9 \times 4$
$A = 36 \text{ m}^2$

8 ft

7 ft.

Rhombus
$A = bh$ (base x height)

$A = 8 \times 7$
$A = 56 \text{ ft}^2$

Parallelogram
$A = bh$ (base x height)

$A = 32 \times 15$
$A = 1820 \text{m}^2$

15 m

32 m

I just mowed 1820 square meters in 91 minutes. That's 20 square meters in a minute. Boy, am I good!

Better Grades & Higher Test Scores / MATH
©Incentive Publications, Inc., Nashville, TN

Triangle

$A = \frac{1}{2} bh$ (½ base × height)

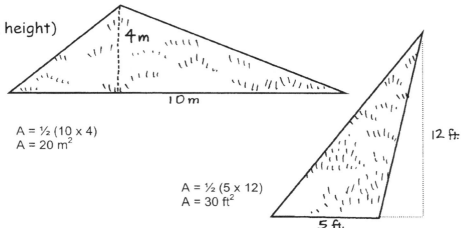

A = ½ (10 × 4)
A = 20 m²

A = ½ (5 × 12)
A = 30 ft²

Trapezoid $A = \frac{1}{2} h(b_1 + b_2)$ (1/2 height × sum of two bases)

A = ½ × 18 × (20 + 15)
A = ½ × 18 × 35
A = 315 yd²

A = ½ × 5 × (20 + 30)
A = ½ × 5 × 50
A = 125 m²

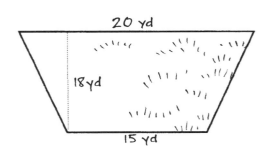

Circle

$A = \pi r^2$ (pi or 3.14 × radius squared)

A = 3.14 × 9²
A = 3.14 × 81
A = 254.34 m²

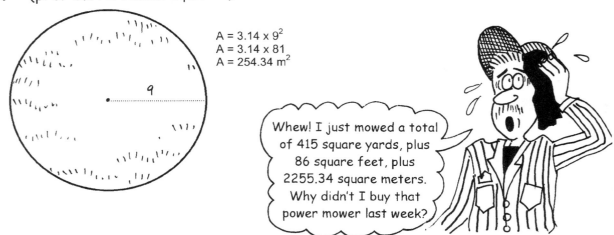

Whew! I just mowed a total of 415 square yards, plus 86 square feet, plus 2255.34 square meters. Why didn't I buy that power mower last week?

Better Grades & Higher Test Scores / MATH
©Incentive Publications, Inc., Nashville, TN

Get Sharp: Measurement

Measuring Surface Area of Space Figures

Surface area is the sum of the areas of all the faces on a space figure.
It is measured in square units.
If you want to decorate, paint, or cover a space figure in any way,
you will need to find the surface area. Fortunately, there are formulas doing this.

cubes

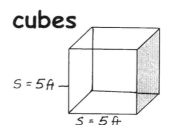

S = sum of 6 faces

6 square faces (s^2)

$S = 6(5^2)$
$S = 6 \times 25$
$S = 150$ ft^2

prisms

S = sum of faces

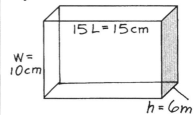

rectangular prism

2 rectangular faces	2 (lw)
2 rectangular faces	2 (lh)
2 rectangular faces	2 (wh)

$S = 2(15 \times 10) + 2(15 \times 6) + 2(10 \times 6)$
$S = (2 \times 150) + (2 \times 90) + (2 \times 60)$
$S = 300 + 180 + 120$
$S = 600$ cm^2

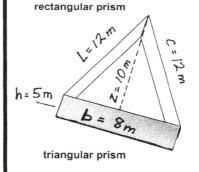

triangular prism

2 triangular faces	$2 \times \frac{1}{2}$(bz)
1 rectangular face	(lh)
1 rectangular face	(hc)
1 rectangular face	(bh)

$S = 2 \times \frac{1}{2}(8 \times 10) + (12 \times 5) + (5 \times 12) + (8 \times 5)$
$S = 2(\frac{1}{2} \times 80) + 60 + 60 + 40$
$S = (2 \times 40) + 160$
$S = 80 + 160$
$S = 240$ m^2

pyramids

S = sum of faces *(number of faces varies depending on base shape)*
b = height of slanted triangular face

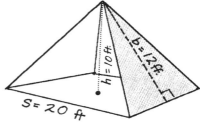

| 1 square face | (s^2) |
| 4 triangular faces | $4 \times \frac{1}{2}$ (s x b) |

$S = 20^2 + 4 (\frac{1}{2} \times 20 \times 12)$
$S = 400 + (4 \times 120)$
$S = 400 + 480$
$S = 880$ ft^2

Get Sharp: Measurement

Better Grades & Higher Test Scores / MATH
©Incentive Publications, Inc., Nashville, TN

spheres

$$S = 4\pi r^2$$

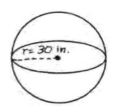

$$S = 4 \times (3.14 \times 30^2)$$
$$S = 4 \times (3.14 \times 900)$$
$$S = 4 \times 2826$$
$$S = 11{,}304 \text{ in}^2$$

cones

$$S = \text{area of base (B)} + \text{area of curved surface}$$

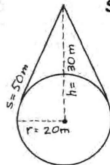

1 circular face (πr^2)
1 curved face (πrs)

$$S = (3.14 \times 20^2) + (3.14 \times 20 \times 50)$$
$$S = (3.14 \times 400) + (3.14 \times 1000)$$
$$S = 1256 + 3140$$
$$S = 4396 \text{ m}^2$$

s = 50m   h = 30 m   r = 20m

cylinders

$$S = \text{area of 2 bases} + \text{area of curved surface}$$

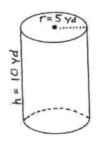

r = 5 yd   h = 10 yd

2 circular faces (πr^2)
1 curved face (πrh)

$$S = 2(3.14 \times 5^2) + (3.14 \times 5 \times 10)$$
$$S = (6.28 \times 25) + (3.14 \times 50)$$
$$S = 157 + 157$$
$$S = 314 \text{ yd}^2$$

I decorated this hat myself. The entire surface area is covered with paint, glitter, and sequins. With a radius of 4 inches and a side-length of 15 inches, the surface area covered was 238.64 square inches.

Better Grades & Higher Test Scores / MATH
©Incentive Publications, Inc., Nashville, TN

Get Sharp: Measurement

Measuring Volume of Space Figures

Volume is the space taken up by an object, or the space inside an object.
Volume is measured in cubic units.

The Rubik's cube is one of the most popular and challenging puzzles of all time.

cubes   V = s³ (s = side)

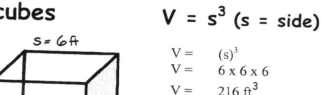

s = 6 ft

$$V = \quad (s)^3$$
$$V = \quad 6 \times 6 \times 6$$
$$V = \quad 216 \text{ ft}^3$$

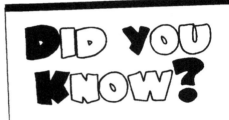

DID YOU KNOW?

A triangular glass prism is a handy tool for learning about light. When light travels through the glass, it is separated into its different wavelengths to show all the colors of the spectrum.

prisms   V = Bh

(B = area of base, h = height)

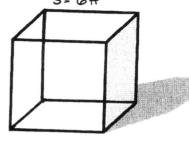

h = 10 cm

L = 15 cm

rectangular prism   w = 6 cm

$$V = \quad (lw)h$$
$$V = \quad (15 \times 6) \times 10$$
$$V = \quad 90 \times 10$$
$$V = \quad 900 \text{ cm}^3$$

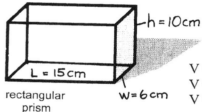

h = 5 m

triangular prism

a = 10

b = 8 m

$$V = \quad (½\,ba)h$$
$$V = \quad (½ \times 8 \times 10) \times 5$$
$$V = \quad (½ \times 80) \times 5$$
$$V = \quad 40 \times 5$$
$$V = \quad 200 \text{ m}^3$$

The Great Pyramid of Giza contains 2,000,000 blocks of stone.

pyramids   V = Bh

$$V = \quad s^2 \times h$$
$$V = \quad (20^2) \times 10$$
$$V = \quad 400 \times 10$$
$$V = \quad 4000 \text{ ft}^3$$

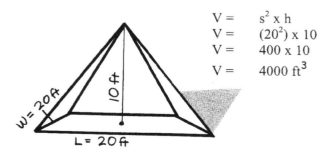

w = 20 ft

10 ft

L = 20 ft

144

spheres $V = \frac{4}{3}\pi r^3$

$$
\begin{aligned}
V &= 4/3 \times (3.14 \times 4.4^3) \\
V &= 4/3 \times (3.14 \times 85.2) \\
V &= 4/3 \times 267.5 \\
V &= 356.6 \text{ in}^3
\end{aligned}
$$

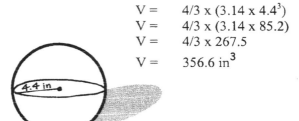

4.4 in

cones $V = \frac{1}{3} Bh$

(B = Area of base)

$$
\begin{aligned}
V &= 1/3 \,(\pi r^2 \times h) \\
V &= 1/3 \,(3.14 \times 20^2 \times 30) \\
V &= 1/3 \,(3.14 \times 400 \times 30) \\
V &= 1/3 \,(1256 \times 30) \\
V &= 1/3 \times 3768 \\
V &= 1256 \text{ cm}^3
\end{aligned}
$$

r = 20cm

h = 30cm

cylinders $V = Bh$

(B = Area of base)

$$
\begin{aligned}
V &= (\pi r^2) \times h \\
V &= (3.14 \times 3^2) \times 12 \\
V &= (3.14 \times 9) \times 12 \\
V &= 28.26 \times 12 \\
V &= 339.12 \text{ cm}^3
\end{aligned}
$$

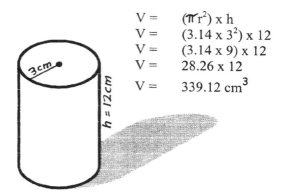

3cm

h = 12cm

Better Grades & Higher Test Scores / MATH
©Incentive Publications, Inc., Nashville, TN

Get Sharp: Measurement

Measuring Capacity

Capacity is the measure of how much of something fits into a container. Capacity may measure liquid (such as water or oil) or dry materials (such as cereal, sugar, raising, oats).
Many different measures are used to find capacity. These get a lot of use in everyday life—in cooking, buying food, picking apples, doing science experiments, and shopping.

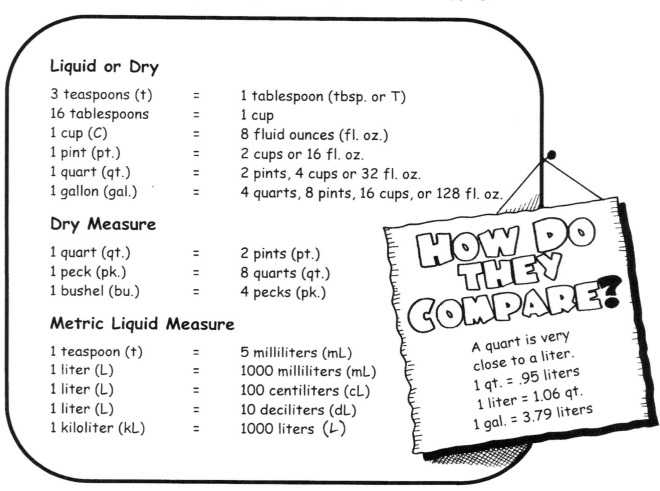

Liquid or Dry

3 teaspoons (t)	=	1 tablespoon (tbsp. or T)
16 tablespoons	=	1 cup
1 cup (C)	=	8 fluid ounces (fl. oz.)
1 pint (pt.)	=	2 cups or 16 fl. oz.
1 quart (qt.)	=	2 pints, 4 cups or 32 fl. oz.
1 gallon (gal.)	=	4 quarts, 8 pints, 16 cups, or 128 fl. oz.

Dry Measure

1 quart (qt.)	=	2 pints (pt.)
1 peck (pk.)	=	8 quarts (qt.)
1 bushel (bu.)	=	4 pecks (pk.)

Metric Liquid Measure

1 teaspoon (t)	=	5 milliliters (mL)
1 liter (L)	=	1000 milliliters (mL)
1 liter (L)	=	100 centiliters (cL)
1 liter (L)	=	10 deciliters (dL)
1 kiloliter (kL)	=	1000 liters (L)

HOW DO THEY COMPARE?

A quart is very close to a liter.
1 qt. = .95 liters
1 liter = 1.06 qt.
1 gal. = 3.79 liters

Measuring the Earth

In order to find locations on Earth, geographers devised a system of measuring the whole surface. Since Earth is a sphere, distance was measured in degrees. (Every circular figure has 360º degrees.)

So Earth is divided into 360º degree-sized sections moving east or west around the sphere. Imaginary lines that extend north and south are called **lines of longitude**. The starting point is the prime meridian at 0º longitude. This line runs through Greenwich, England. Longitude is measured in units east or west of the º0 longitude. Longitude gives a location as distance east or west of the prime meridian. Lines are numbered from 0º to 180 ºat the International Date Line.

Other imaginary lines circle the globe from North Pole to South Pole. These are **lines of latitude**. The equator circles Earth at its widest circumference. The equator is the 0º line of latitude. Latitude is measured in segments north or south of the equator. Lines of latitude go from 0 at the equator to 90º N at the North Pole and 90ºS at the South Pole.

Locations on Earth are given in terms of their latitude and longitude. Locations are written like this:

55º N, 17º E. *(latitude, longitude)*

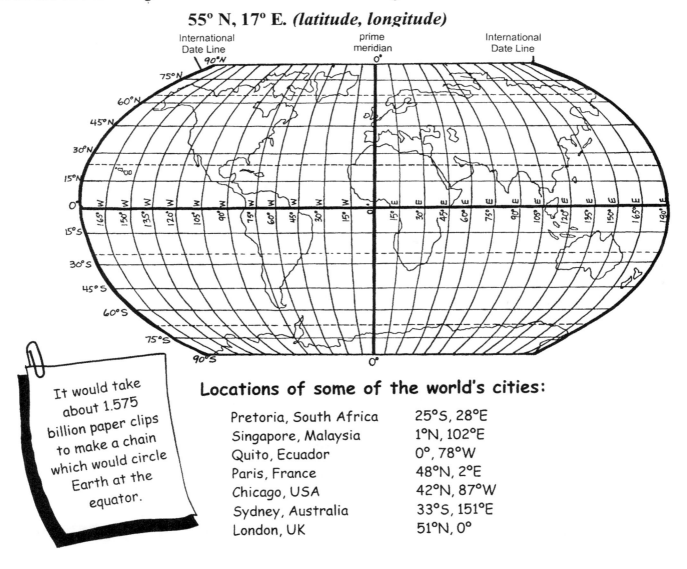

Locations of some of the world's cities:

It would take about 1.575 billion paper clips to make a chain which would circle Earth at the equator.

Pretoria, South Africa	25ºS, 28ºE
Singapore, Malaysia	1ºN, 102ºE
Quito, Ecuador	0º, 78ºW
Paris, France	48ºN, 2ºE
Chicago, USA	42ºN, 87ºW
Sydney, Australia	33ºS, 151ºE
London, UK	51ºN, 0º

Measuring Time

The measurement of time is based on the revolution of Earth. This Earth movement takes 24 hours and causes night and day.

The period of the revolution is called a **day**. Time has been further divided into other segments.

Calendars, clocks, and watches are the most common tools for measuring time.

60 seconds (sec)	=	1 minute (min)
60 minutes (min)	=	1 hour (hr)
24 hours (hr)	=	1 day
7 days	=	1 week
2 weeks	=	1 fortnight
4 weeks (approx)	=	1 month
365-366 days	=	1 year
12 months	=	1 year
10 years	=	1 decade
100 years	=	1 century
1000 years	=	1 millennium

Standard Time

In the USA, the measure used is Standard Time. This measures the day in two 12-hour blocks—before noon and after noon.

The first 12 hours begins at midnight and the hours are numbered 1-12 up to noon. This is called a.m. for *ante meridiem*, meaning *before noon*.

The second block begins at noon and numbers the hours 1-12 until midnight. This is called p.m. for *post meridiem*, meaning *after noon*.

Military Time

The military and most other countries use a time measurement system that labels hours 1-24 beginning at midnight. That way, when you say 9 o'clock, no one has to wonder if you mean 9 in the morning or evening.

In military time, the time is described in hundreds of hours.

9:00 (morning)   = 900 hours

2:00 (afternoon) = 1400 hours

11:00 (evening) = 2300 hours

Better Grades & Higher Test Scores / MATH
©Incentive Publications, Inc., Nashville, TN

As Earth turns, the sun rises and shines on different parts of the sphere at different times. Because of this, Earth is divided into several time zones.

The line of 0° longitude (the Prime Meridian) goes through Greenwich, England. Earth's time zones are all related to the time in Greenwich, called Greenwich time. From Greenwich, an hour is subtracted as you travel west through each time zone. As you travel east from Greenwich, an hour is added to the time.

When it's noon in Greenwich Village, it's time for lunch!

When it is noon in Greenwich, England. . .

It is 7 a.m. (5 time zones west)	in New York City, NY
it is 6 a.m. (6 time zones west)	in Dallas, TX
it is 4 a.m. (8 time zones west)	in Los Angeles, CA
it is 3 a.m. (9 time zones west)	in Anchorage, AK
it is 1 p.m. (1 time zone east)	in Paris, France
it is 3 p.m. (3 time zones east)	in Moscow, Russia
it is 8 p.m. (8 time zones east)	in Hong Kong
it is 9 p.m. (9 time zones east)	in Tokyo, Japan

and . . . it is midnight (12 time zones east or west) on the International Date Line (180° E or W longitude)

Measuring Rate

Rate is a measure of the amount of distance covered over time, or the amount of some other event. Rate tells how far something moves or how frequently something occurs over a specific period of time, such as a second, minute, hour, week, year, and so on. Speed is always described as a rate.

— SOME INTERESTING RATES —

186,282,397 miles per second (mps)	speed of light
66 miles per hour (mph)	speed a sailfish can swim
65 kilometers per hour (kph)	speed a mallard duck can fly
12 miles per hour (mph)	speed of a running rabbit
11.6 kilometers per hour (kph)	speed a honeybee can fly
0.03 miles per hour (mph)	speed a snail can crawl
261.8 miles per hour (mph)	speed of Japan's fast Nozomi 500 train
29.2 pints per year	amount of ice cream eaten by average Australian
26 pounds per year	amount of chocolate eaten by average Swiss person
48 gallons per year	amount of soda pop drunk by average American
100 miles per hour (mph)	speed of world's two fastest roller coasters *(Tower of Terror in Australia and Superman: The Escape in Valencia, CA)*

Measuring Temperature

Temperature is measured to learn about the heat in some substance.
(Many times the substance is the air!)
There are two scales commonly used for measuring temperature.

Get Sharp Tip # 22
Use this formula to convert Fahrenheit temperatures to Celsius:

$$\frac{(F - 32)\ 5}{9}$$

The Fahrenheit Scale

A German scientist, Gabriel Fahrenheit, developed a scale for measuring heat. This is the scale most frequently used in the United States. The Fahrenheit scale is based on the temperature at which water freezes and boils. 0° Fahrenheit is the temperature of a mixture of solid water (snow) and ice. 32° is the temperature at which water freezes. 212° is the temperature at which water boils.

The Celsius Scale

Swedish astronomer, Anders Celsius, developed a different scale for measuring heat. The Celsius (or Centigrade) scale is used by scientists and in most countries other than the US. This scale is also based on the freezing and boiling points. 0° Celsius is the temperature at which water freezes. The boiling point of water is 100° C.

Get Sharp Tip # 23
Use this formula to convert Celsius temperatures to Fahrenheit:

$$\frac{9C + 32}{5}$$

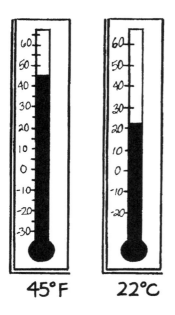

45°F   22°C

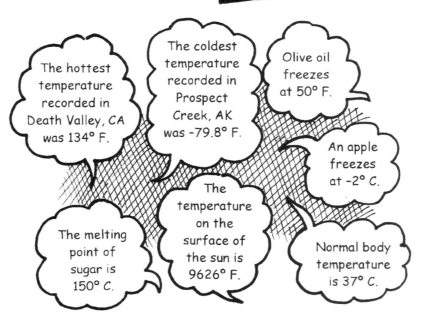

The hottest temperature recorded in Death Valley, CA was 134° F.

The coldest temperature recorded in Prospect Creek, AK was -79.8° F.

Olive oil freezes at 50° F.

An apple freezes at -2° C.

The melting point of sugar is 150° C.

The temperature on the surface of the sun is 9626° F.

Normal body temperature is 37° C.

More Measures

There are many more units used to measure many different things in our world.

Here are a few of them. ➡

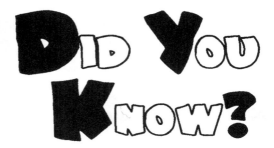

- The *Titanic* was traveling at its top speed of 21 knots when icebergs were sighted.

- The wreckage of the *Titanic* lies about 347 nautical miles southeast of Newfoundland, at a depth of about 2000 fathoms.

- The deepest spot in any ocean, the Mariana Trench, is 5973 fathoms deep.

- The star, Orion, is 900 light years from Earth.

- Earth is 1 AU from the sun.

- Jupiter is 5.2 AU from the sun.

- The largest diamond ever found, the Cullinan Diamond, weighs 3106 carats. This diamond, now in two parts, is part of the British crown jewels.

- The Hope Diamond weighs 45.52 carats.

fathom - a unit measuring ocean depth

= 6 feet

cable - a unit measuring ocean depth

= 120 fathoms or 720 feet

nautical league - a unit measuring distance at sea, usually of ocean depth

= 3 nautical miles or 5.6 kilometers

nautical mile - a unit measuring distance at sea

= 6076.1 feet

knot - a unit measuring speed at sea

= 1 nautical mile per hour

light year - a unit measuring time in space

= distance light travels in a year, 5878 trillion miles or 9.5 trillion km

AU (Astronomical Unit) - a unit measuring distance in the solar system, based on the distance from Earth to sun

1 AU = 92.9 million mi.

grain - a unit measuring weight, originally based on the weight of a grain of wheat

= 0.002083 oz.

carat - a unit measuring weight of precious metals and gems

= 3.086 grains

decibel - a unit measuring the intensity of sound

Normal conservation is about 60 decibels.

The softest sounds humans can hear measure 0 decibels. Loud rock music is about 120 decibels.

Blue whales make a sound of 188 decibels.

Can dogs hear blue whales?

hand - a unit measuring the height of horses

= the approximate width of an adult hand

calorie - a unit measuring heat energy

= the amount of heat needed to raise the temperature of one gram of water 1° C

joule - a unit measuring heat energy

= 0.239 calories

BTU (British Thermal Unit) - a unit measuring heat

= 251.996 calories

Newton - a unit measuring force

= the amount of force required to increase or decrease the velocity of a one kilogram weight by one meter per second every second

volt - a unit measuring the ability of electricity to give energy

= enough energy to push 1 ampere of current through an electrical resistance of 1 ohm (a measure of electrical resistance)

ampere - a unit measuring the rate of flow of electrical current

= one unit of electric charge flows past a point in an electrical circuit in one second

foot-pound - a unit measuring power

1 foot-pound = the amount of power needed to move one pound weight one foot

horsepower – a unit measuring the rate of doing a work

1 horsepower = 550 foot-pounds per second or 33,000 foot-pounds per minute

Get Sharp Tip # 24
Joules are also used to measure work.

GET SHARP →

on

Statistics, Graphing, & Probability

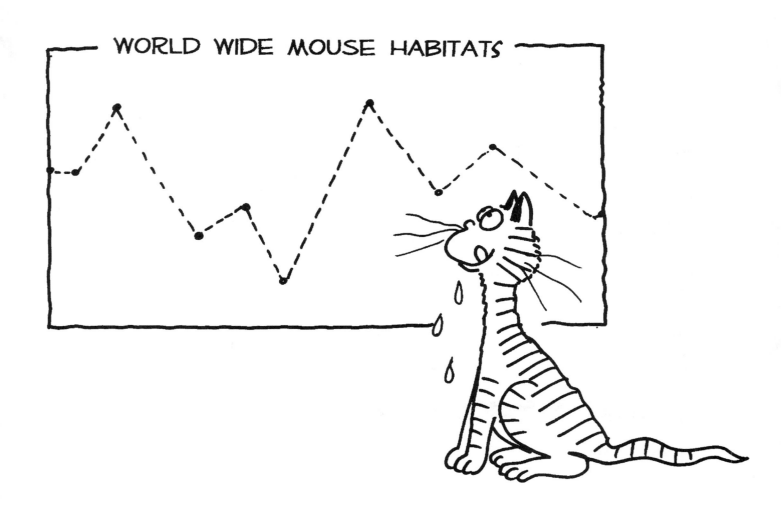

WORLD WIDE MOUSE HABITATS

Statistics

Statistics is a branch of mathematics that deals with numerical information (called **data**). Statistics involves collecting, organizing, presenting, and interpreting data.

Many times, a purpose for collecting data on a certain topic is to be able to compare the numbers.

A table is one common way of organizing and presenting data.

> I walked 870 miles on my hands. . .

Records for Tightrope Walks
Mile High County Tightrope Club

Tightrope Walker	Record Height of Wire (in feet)	Record Length of Walk (in feet)
Francine	430	48
Flossie	610	45
Frankie	225	30
Frenchie	80	55
Phillipe	610	35
Flo	390	72
Phyllis	850	29
Flip	1000	120
Fran	305	85
Phoebe	700	29
Frank	903	68
Fred	275	72

RECORDS for EXTREME JOURNEYS
(According to 1999 Guinness Book of World Records, Rounded to the nearest mile)

Journey	Distance in Miles	Journey	Distance in Miles
Taxi	21,691	Polar Sled	3,750
Motorcycle	457,000	Bicycle	226,800
Snowmobile	10,252	Backwards Walk	8,000
Lawn Mower	3,366	Backwards Run	3,100
Wheelchair	24,903	Hitchhike	501,750
Unicycle	2,361	Stilt Walk	3,008
Walk on Hands	870	Walk on Water	3,502
Leapfrog	996	Parachute Fall	6
Skates	19,000	Unicycle, riding backwards	53
Crawling	870	Dancing	23

Get Sharp Tip # 25

Make sure any table of data you create has a clear title and clear labels for all rows and columns.

> . . . is there a statistic about the number of blisters on the hands of hand walkers?

Range, Mean, Median, & Mode

If you're going to understand statistics, *range, median, mode* and *mean* are all terms you should know. They are some of the most helpful words in statistics, because they help to describe sets of data.

Number of Flat Tires During Bike Marathon									
Biker's Name	A.J. Ryder	J.R. Crash	Tom Elite	Gabe McTrick	Abby deWheel	B.B. Wynn	Z.Z. Tubes	Flip Slykes	C.C. Cross
Number of flat tires over the 2-week race	13	6	14	21	18	3	7	7	10

RANGE is the difference between the least and the greatest numbers in the set of data.

The range here is

3-21 flat tires.

MEAN is the average of the data.
To find the mean, divide the sum of all the data by the number of data items.
For this data, the mean =

$$\frac{13 + 6 + 14 + 21 + 18 + 3 + 7 + 7 + 10}{9} = \frac{99}{9}$$

The mean is 11.

MEDIAN is the number in the middle of a set of data.
The numbers in this set are:

3-6-7-7-10-13-14-18-21

The median is 10.

MODE is the number that appears most often in a set of data.

The mode here is 7.

(Sometimes there is no mode. Sometimes there are two or more modes.)

Get Sharp Tip # 26

To find the median, first arrange all the data items into numerical order.

Better Grades & Higher Test Scores / MATH
©Incentive Publications, Inc., Nashville, TN

Get Sharp: Statistics

Frequency

The **frequency** of a number means how often it appears in a set of data. Sometimes it is useful or necessary to find out how often a number (or group of numbers) shows up in a set of data.

This set of data shows the ages of all the competitors who registered for a bull-riding event in a rodeo. As each rider registered, his or her age was written on a list.

To find out how many of the riders fell into certain age groups, the data was organized into a frequency table.

1. First the ages were grouped into intervals (15-24, 25-34, and so on).

2. Next, a tally mark was placed into the correct tally column each time an age was recorded.

3. Finally, the tally marks were counted.

4. The frequency of ages in each group was written as a number in the *frequency* column.

Ages of Bull-Riding Competitors

26	22	16	56	40	36
68	20	17	20	41	47
15	18	23	18	26	19
32	46	29	60	17	56
51	66	35	28	29	24
40	76	49	33	68	34
30	77	50	66	31	30
16	16	20	26	31	23
20	40				

Ages of Bull-Riding Competitors

Ages	Tally	Frequency
15-24	ʬ ʬ ʬ ‖	17
25-34	ʬ ʬ ⦀	13
35-44	ʬ ∣	6
45-54	ʬ	5
55-64	⦀	3
65-75	⦀⦀	4
over 75	‖	2

The age of the bull riders really varies!

Yeah, but how old are the bulls?

Get Sharp Tip # 27

Relative frequency is

$$\frac{\text{frequency of an item}}{\text{total of frequencies}}$$

The relative frequency of riders over 75 years old is 2:50 or 2/50.

Better Grades & Higher Test Scores / MATH
©Incentive Publications, Inc., Nashville, TN

Bar Graphs

A bar graph uses bars of different lengths to show and compare data.
A single bar graph shows one kind of data. This single bar graph shows the number of times
Ollie practiced each wakeboard trick in a week.

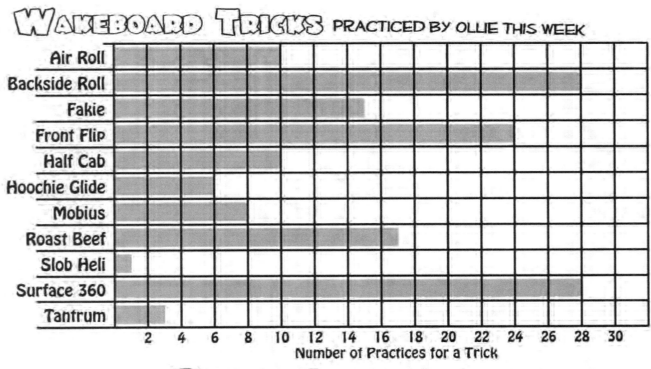

WAKEBOARD TRICKS PRACTICED BY OLLIE THIS WEEK

Number of Practices for a Trick

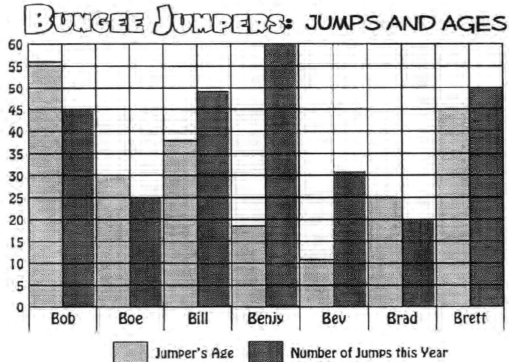

BUNGEE JUMPERS: JUMPS AND AGES

Jumper's Age   Number of Jumps this Year

Histograms

A **histogram** is a bar graph that shows frequency data. This table shows injuries from a snowboarding competition. The data from the table has been displayed below in graph form.

From reading the graph, you can see the frequency with which each type of injury occurred.

INJURIES

Kind of Injury	Number
Broken Legs	1
Broken Arms	2
Broken Feet	3
Broken Wrists or Fingers	6
Broken Noses	3
Broken Teeth	17
Sprained Ankles	5
Sprained Wrists or Fingers	28
Pulled Ligaments	23
Back Injuries	12
Head Injuries	5
Shoulder Injuries	25
Serious Cuts	32
Bloody Noses	26
Frostbite Cases	13
Serious Sunburn	22

As you can see on the graph, my injury is one of the most common!

SEASON-LONG SNOWBOARDING INJURIES

Number of Injuries

15,000 spectators watched Kelly Clark win the women's snowboarding halfpipe competition at the 2002 Winter Olympics.

After watching the 2002 Olympic snowboard competitions, 18.6 million people said they would like to try the sport.

Circle Graphs

A circle graph pictures the sizes or amounts of data items as they join together in a total amount. A circle graph makes it possible to see what part of a whole each data item represents. Circle graphs are sometimes called *pie graphs* or *pie charts*.

This circle graph displays the costs of the scuba gear. Each segment of the graph shows what part of the total cost is used up by each individual piece of gear.

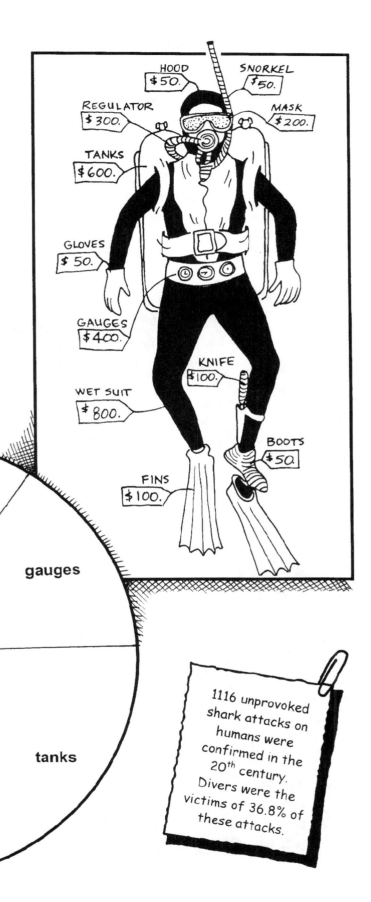

HOOD $50.
SNORKEL $50.
REGULATOR $300.
MASK $200.
TANKS $600.
GLOVES $50.
GAUGES $400.
KNIFE $100.
WET SUIT $800.
BOOTS $50.
FINS $100.

The circle graph segments: mask, regulator, gauges, tanks, wet suit, boots, gloves, snorkel, hood, knife, fins

1116 unprovoked shark attacks on humans were confirmed in the 20th century. Divers were the victims of 36.8% of these attacks.

Better Grades & Higher Test Scores / MATH
©Incentive Publications, Inc., Nashville, TN

Get Sharp: Graphing & Statistics

Pictographs

A **pictograph** uses pictures, symbols, or icons to display and compare data. Each picture represents a specific data number. Fractions of the picture may be used to represent fractions of the data amount. A key shows what the picture represents.

> After I eat this pile of pancakes, I'll wait at least an hour before going water-skiing.

> Ralf Laue of Germany holds the World Pancake-Tossing record. He tossed 416 pancakes in 3 minutes.

Pancakes Eaten at Annual Pancake-Eating Contest

 = 100 pancakes

Team A *The Syrup Kings*	
Team B *The Gobblers*	
Team C *The Flapjack 5*	
Team D *The Munchies*	
Team E *The Tall Stack Guys*	

Get Sharp: Graphing & Statistics

Line Graphs

A **line graph** uses a line on a grid to show data over time. A line graph has a special talent that other kinds of graphs don't have: it is able to show changes in data over a period of time.

This graph shows the amount of time Dixie practiced her water-skiing tricks over several weeks.

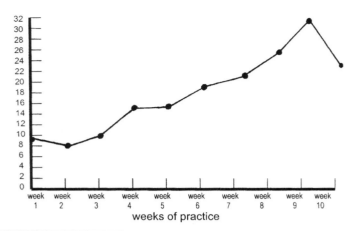

Dixie's Practice Time

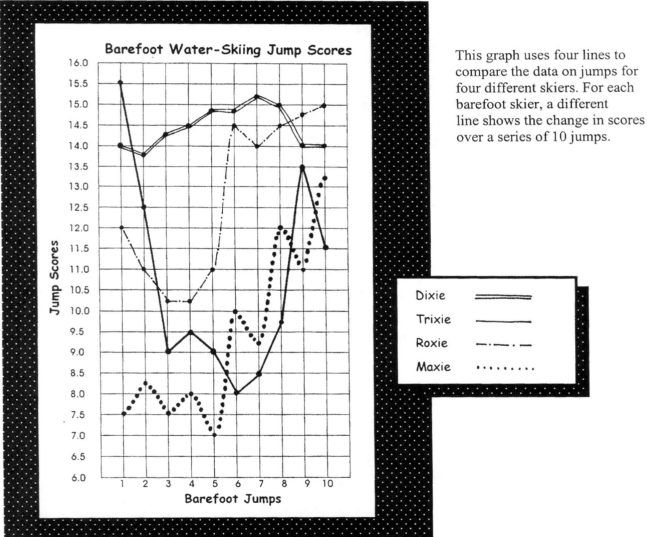

This graph uses four lines to compare the data on jumps for four different skiers. For each barefoot skier, a different line shows the change in scores over a series of 10 jumps.

Scattergrams

A **scattergram** is a type of graph that uses a grid
to show the relationships between two quantities.

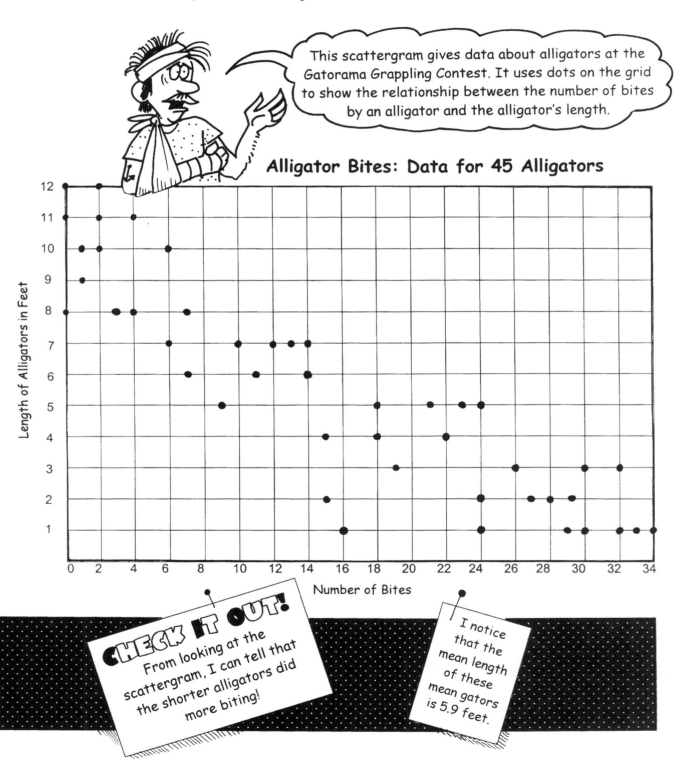

This scattergram gives data about alligators at the
Gatorama Grappling Contest. It uses dots on the grid
to show the relationship between the number of bites
by an alligator and the alligator's length.

Alligator Bites: Data for 45 Alligators

Length of Alligators in Feet

Number of Bites

CHECK IT OUT! From looking at the scattergram, I can tell that the shorter alligators did more biting!

I notice that the mean length of these mean gators is 5.9 feet.

What is Probability?

Probability is the chance or likelihood that something will happen. A number from zero to one (0 to 1) is used to describe the probability of something happening.

A probability of **0** means something **will not** happen.

A probability of **1** means something **is certain to happen**.

Probability is expressed as 0, 1, or as a ratio, fraction, or percentage.

Ceci has four marbles in her pocket: one yellow, one black, and two red.

If Ceci reaches into her pocket without looking and grabs one marble, the probability that she'll choose a yellow marble is one in four.

This can be expressed 1:4, $\frac{1}{4}$, or 25%.

Write $P(Y) = \frac{1}{4}$

Meaning: The probability of yellow is 1 in 4.

I can't remember how many marbles I **do** have! Have I lost my marbles?

- The probability that the sun will set in the west is **1**.

- The probability that a random day picked will be a Friday is $\frac{1}{7}$.

- The probability that fall will follow winter is **0**.

- The probability that two odd numbers will have an even sum is **1**.

- The probability that the toss of 1 die will yield 5 is $\frac{1}{6}$.

- The probability that Ceci will choose a red marble is $\frac{2}{4}$.

- The probability that the toss of 1 coin will yield *heads* is $\frac{1}{2}$.

- The probability that one spin of this spinner will yield green is $\frac{1}{3}$.

- The probability that one spin of this spinner will not yield green is $\frac{2}{3}$.

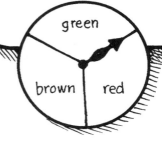

Outcomes & Events

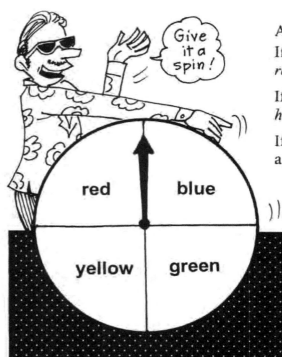

Give it a spin!

An **outcome** is the result of an action or experiment. If you spin this spinner, there are four possible outcomes: *red, blue, yellow,* or *green.*

If you flip a coin, the number of possible outcomes is two: *heads* or *tails.*

If you toss one die, the number of possible outcomes is six: a *1, 2, 3, 4, 5,* or *6.*

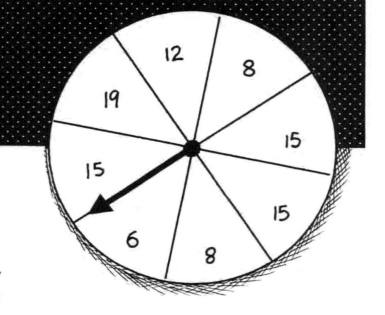

An **event** is a particular outcome or set of outcomes. If you spin this spinner there are eight possible outcomes. However, there are three chances that the event could be 15.

For this spinner, some events are more likely to happen than others. The event of stopping on 15 is certainly more likely than the event of stopping on 6.

$$\text{Probability of an event} = \frac{\text{the number of possible events}}{\text{the total number of possible outcomes}}$$

The probability of stopping on 15 is written this way: **P(15)**

$$P(15) = \frac{\text{number of 15s}}{\text{total possible outcomes}} = \frac{3}{8}$$

Other probabilities with this spinner:

$$P(8) = \frac{\text{number of 8s}}{\text{total outcomes}} = \frac{2}{8}$$

$$P(15 \text{ or } 8) = \frac{\text{number of 15s \& 8s}}{\text{total outcomes}} = \frac{5}{8}$$

$$P(<10) = \frac{\text{number of outcomes} <10}{\text{total outcomes}} = \frac{3}{8}$$

Outcomes of Two Actions

When two actions happen, the possible outcomes can be shown on a table.

Two coins are tossed. . .

The table shows the four possible outcomes. What is the probability that both coins will land with *heads* facing up?

To find the probability, use this formula:

Probability (of an event) = $\dfrac{\text{the number of possible events}}{\text{the total number of possible outcomes}}$

P (H, H) = ¼
P (both tosses will yield the same result) = 2/4

Two dice are tossed. . .

The table shows the 36 possible outcomes.

P(6, 6)   = 1/36

P both the same)   = 6/36 (or 1/6)

P(a 3 and a 4)   = 2/36 (or 1/18)

P (both numbers > 3) = 9/36 or ¼

Jojo spins both spinners once. . .

The table shows the 12 possible outcomes.

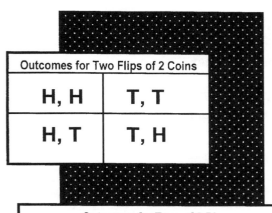

Outcomes for Two Flips of 2 Coins	
H, H	**T, T**
H, T	**T, H**

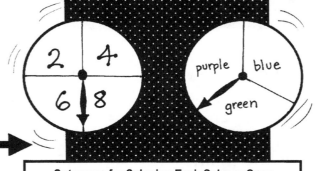

Outcomes for Toss of 2 Dice					
1,1	2,1	3,1	4,1	5,1	6,1
1,2	2,2	3,2	4,2	5,2	6,2
1,3	2,3	3,3	4,3	5,3	6,3
1,4	2,4	3,4	4,4	5,4	6,4
1,5	2,5	3,5	4,5	5,5	6,5
1,6	2,6	3,6	4,6	5,6	6,6

Outcomes for Spinning Each Spinner Once			
2, b	4, b	6, b	8, b
2, p	4, p	6, p	8, p
2, g	4, g	6, g	8, g

Probability of Independent Events

Events are independent when the outcomes of one event is not affected by the outcome of the other.

If a coin is flipped and one die is tossed, the outcome of the coin toss will not affect the outcome of the toss of the die.

To find the probability of two independent events, multiply the probability of one event by the probability of the other.

P(\$100 and red)   =   P(100) x P(R) = 1/6 x 2/5 =   2/30 or 1/15

P(nor \$50 and not red) =   P(not 50) x P(not R) = 5/6 x 3/5 = 15/30 or 1/2

P(\$5 and pink)   =   P(5) x P(P) = 1/6 x 1/5 = 1/30

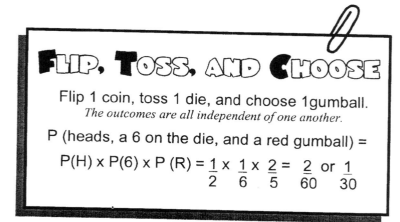

FLIP, TOSS, AND CHOOSE

Flip 1 coin, toss 1 die, and choose 1 gumball.
The outcomes are all independent of one another.

P (heads, a 6 on the die, and a red gumball) =

$P(H) \times P(6) \times P(R) = \frac{1}{2} \times \frac{1}{6} \times \frac{2}{5} = \frac{2}{60}$ or $\frac{1}{30}$

The coins called nickels made from 1942-1945 are not nickel at all! They are 56% copper, 35% silver, and 9% manganese.

Probability of Dependent Events

Events are dependent when the outcome of one event
is affected by the outcome of a previous event.

Get Sharp
Tip # 28

Why are Abby's
choices of socks
dependent events?
Well, after she
chooses one sock,
there are no longer
12 socks, and the
remaining color choices
have changed!

- For instance, Abby has 12 socks in her drawer: 3 blue, 4 white, 2 red, and 3 pink. She reaches in without looking and pulls out two socks, one at a time. The probability of her second choice will be affected by the first choice.

- The probability of one event (B) happening, given that another event (A) has already taken place is written like this:

    P(B|A) and reads *the probability of B given A.*

- Use this formula for the probability of two dependent events:

    P(A and B) = P(A) x P(B|A)

- Abby pulls out a blue sock first, and reaches in for another.

    *What is the probability that the second sock will be **white**?*

    $$P(B \text{ and } W) = P(B) \times P(w/B)$$
    $$= \frac{3}{12} \times \frac{4}{11} = \frac{12}{132} = \frac{1}{11}$$

- Abby's first draw is pink.

    *What is the probability that the second sock will be **blue**?*

    $$P(P \text{ and } B) = P(P) \times P(B/P)$$
    $$= \frac{3}{12} \times \frac{3}{11} = \frac{9}{132} = \frac{3}{44}$$

- *What is the probability she will draw out a **red pair** on the first two draws?*

    $$P(R \text{ and } R) = P(R) \times P(R/R)$$
    $$= \frac{2}{12} \times \frac{1}{11} = \frac{2}{132} = \frac{1}{66}$$

Tree Diagrams

A tree diagram is an interesting and helpful visual tool for figuring probability. All the possible outcomes for events can be shown on a tree diagram.

Get Sharp Tip # 29

A tree diagram is a good tool to use to show outcomes for two independent events.

In a 2-day game, Reggie has a chance of winning great prizes. The first day, he can draw one of four envelopes. Two of the envelopes each hold $100, the third envelope holds $50, and the fourth envelope holds $20.

On the second day, Reggie has an equal chance of winning one of these prizes: a CD player (C), a skateboard (S), or a movie pass (M).

The tree diagram shows the possible outcomes for the 2-day game for Reggie.

The Tree Diagram Shows Possible Outcomes

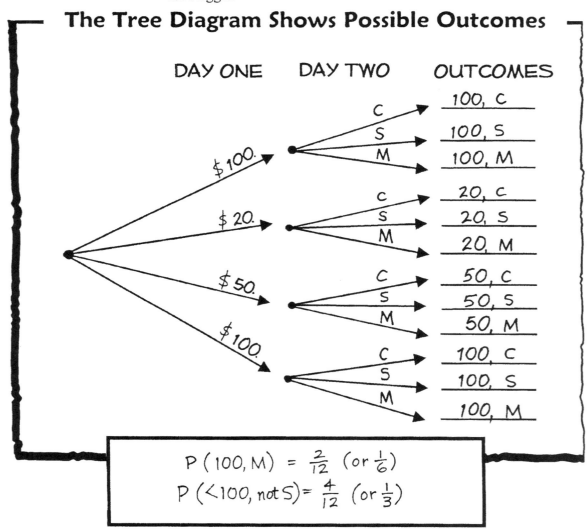

DAY ONE   DAY TWO   OUTCOMES

$100. → C → 100, C
       → S → 100, S
       → M → 100, M

$20. → C → 20, C
      → S → 20, S
      → M → 20, M

$50. → C → 50, C
      → S → 50, S
      → M → 50, M

$100. → C → 100, C
       → S → 100, S
       → M → 100, M

$$P(100, M) = \frac{2}{12} \ \left(or \ \frac{1}{6}\right)$$

$$P(<100, not \ S) = \frac{4}{12} \ \left(or \ \frac{1}{3}\right)$$

The Counting Principle

Sometimes the outcomes of actions are so numerous or complicated that it doesn't make sense to list them all on a table or show them all on a tree diagram. Usually this is true when there are several stages or parts to an event. In these cases, you can find the number of possible outcomes by using the **counting principle**.

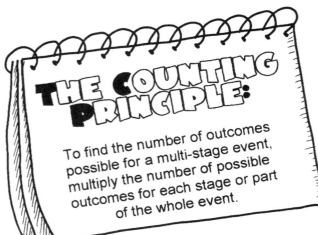

THE COUNTING PRINCIPLE:

To find the number of outcomes possible for a multi-stage event, multiply the number of possible outcomes for each stage or part of the whole event.

A skier tries to decide which route to take down the ski hill. How many choices are there, exactly, on each hill?

Multiply the possible routes for each section. The final product will show how many possibilities there are for each hill.

I tried 156 different routes!

Hill Number 1

How many possibilities are there from start (A) to finish (D)?

3 x 3 x 4 = 36

Hill Number 2

How many possibilities are there from start (1) to finish (5)?

4 x 2 x 3 x 3 = 72

Hill Number 3

How many possibilities are there from start (W) to finish (Z)?

4 x 4 x 3 = 48

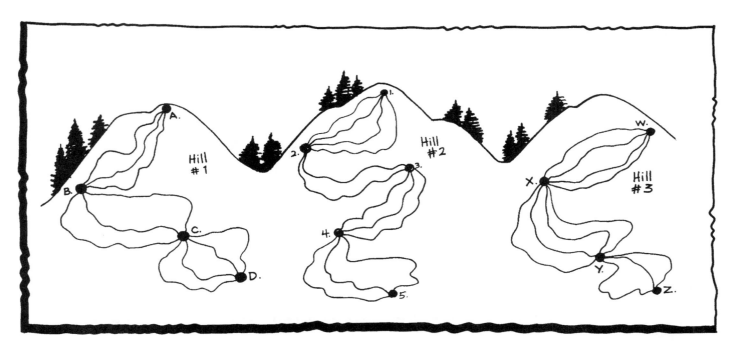

Better Grades & Higher Test Scores / MATH
©Incentive Publications, Inc., Nashville, TN

Get Sharp: Probability

Permutations & Combinations

A **permutation** is a different arrangement of events or items.

A rat, a cat, and an ostrich raced a mile. There are several different ways these three runners could finish the race in 1st, 2nd, and 3rd places.

These possibilities (permutations) can be shown on a table. Or, you can use a formula to find the number:

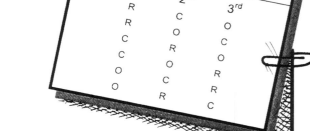

Permutations for 3 Racers		
R = rat, C = cat, O = ostrich		
1st	2nd	3rd
R	C	O
R	O	C
C	R	O
C	O	R
O	R	C
O	C	R

3 choices for 1st place x 2 choices for 2nd place x 1 choice for 3rd place

3 x 2 x 1 = 6 permutations

The number of permutations for . . .

. . . 5 friends in line for tickets $= 5 \times 4 \times 3 \times 2 \times 1 = 120$

. . . 4 books on a bookshelf $= 4 \times 3 \times 2 \times 1 = 24$

. . . 6 pictures on a wall $= 6 \times 5 \times 4 \times 3 \times 2 \times 1 = 720$

A **combination** is a set of things made from a larger set. In a combination, the order does not make a difference.

Ceci has four warm-up shirts in her closet: 1 white, 1 red, 1 green, and 1 blue. She wants to take 3 of the 4 on a trip.

What are the possible combinations she can pack?

Note: Some combinations will be the same. For instance, red and blue is the same as blue and red.

Cross out repeating combinations.

There are 6 different combinations possible.

Combinations for 3 of 4 shirts

WR   RG   GB
WG   RB   B̶W̶
WB   G̶W̶   B̶R̶
R̶W̶   G̶R̶   B̶G̶

Odds

Odds is a term used to describe the chance that something can happen. It is different from probability, in that it compares the number of favorable outcomes to the number of unfavorable outcomes.

Odds are written as a fraction or a ratio.

I have been told that there is a great prize hidden under one of four boxes. If I choose one, what are the odds that I will get the prize?

Odds in Favor

Maxie's odds **in favor** of getting the prize are 1 to 3.
This is the ratio:

$$\frac{\text{number of favorable outcomes}}{\text{number of unfavorable outcomes}} = \frac{1}{3}$$

or   1 to 3

Odds Against

Maxie's odds **against** getting the prize are 3 to 1.
This is the ratio:

$$\frac{\text{number of unfavorable outcomes}}{\text{number of favorable outcomes}} = \frac{3}{1}$$

or   3 to 1

   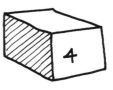

If you know the odds, you can find the probability.

9 boxes; 8 contain a prize
Choose 1 box.

odds in favor of getting a prize	= 8/1
odds against getting a prize	= 1/8
probability of getting a prize	= 8/9

6 boxes; 2 contain a prize
Choose 1 box.

odds in favor of getting a prize	= 2/4
odds against getting a prize	= 4/2
probability of getting a prize	= 2/6

Better Grades & Higher Test Scores / MATH
©Incentive Publications, Inc., Nashville, TN

Sampling

Sampling is a method of estimating or predicting an approximate number of events that will occur among a very large set of items or components.

The idea of sampling is that a small number of items from a larger group are chosen at random and examined. (This is easier than looking at every item in a large group.) The statistics from the small sample are then used to predict what would be true of the whole group.

A sampling of 50 passengers on an airplane found that 6 never had measles. I predict that, of the 350 passengers on the plane, 42 had never had measles.

I am a whiz at candy sampling!

A light bulb company manufactures 5000 light bulbs a week. Unfortunately, not all of them are perfect. Some don't work! To find out how many light bulbs are defective each day, the manager took out a group of 100 light bulbs and tested them. He found 7 light bulbs defective.

He used this proportion to predict how many light bulbs would be defective in the batch of 5000.

$$\frac{\text{number of defective bulbs in the sample}}{\text{number in the random sample}} = \frac{\text{total number of defects}}{\text{total number of bulbs}}$$

$$\frac{7}{100} = \frac{x}{5000}$$

$$100\,x = 35{,}000$$

$$x = 35{,}000 \div 100$$

$$x = 350 \text{ defective bulbs}$$

This bag of taffy has 1200 pieces! I've done a sampling by randomly drawing 40 pieces and eating them. I have eaten 12 chocolate, 5 banana, 8 peppermint and 15 peanut butter. I can predict the total number of each flavor that were in the bag: 360 chocolate, 150 banana, 240 peppermint, and 450 peanut butter.

Will you share?

GET SHARP →

on

Pre-Algebra

Integers

The set of **integers** is
the set of numbers including zero
and all numbers greater than or less than zero.

Positive integers are greater than zero.
Negative integers are less than zero.

Do you have to do everything exactly opposite from what I do?

Absolutely!

The number line is a graph of some points corresponding to integers.
The location of each point is called the **coordinate**.

A is the graph of negative 6.   Its coordinate is –6

B is the graph of negative 2.   Its coordinate is –2.

C is the graph of positive 2.5.   Its coordinate is 2.5.

D is the graph of positive 7.   Its coordinate is 7.

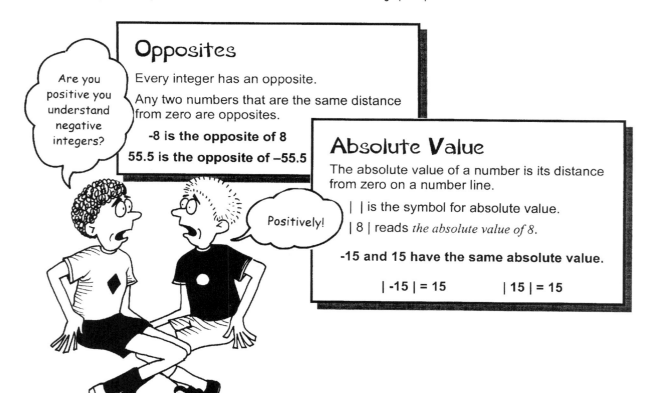

Are you positive you understand negative integers?

Opposites

Every integer has an opposite.

Any two numbers that are the same distance from zero are opposites.

-8 is the opposite of 8

55.5 is the opposite of –55.5

Positively!

Absolute Value

The absolute value of a number is its distance from zero on a number line.

| | is the symbol for absolute value.

| 8 | reads *the absolute value of 8.*

-15 and 15 have the same absolute value.

| -15 | = 15   **| 15 | = 15**

Get Sharp: Integers

Better Grades & Higher Test Scores / MATH
©Incentive Publications, Inc., Nashville, TN

Using Integers

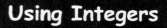

A diver watches fish at a depth of 40 meters. The integer is
-40.

Last night the temperature dropped to 21° below zero. The integer is
-21.

A climber scales the side of a 285-foot building. The integer is
285.

Joey's bank account is overdrawn by $75. The integer is
-75.

Shaun's temperature rose to 103.5°. The integer is
103.5°.

Comparing Integers

All positive numbers are greater than all negative numbers.

To compare integers, picture their location on a number line.

-12 > -15

-20 < 2

1.7 > -7

-13 < -9 > 5 reads:

−9 is between −13 and 5

or *−9 is greater than −13 and less than 5*

Ordering Integers

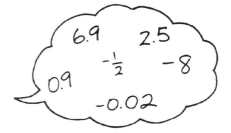

6.9   2.5 $-\frac{1}{2}$ -8   0.9   -0.02

To order integers, start with the number with the least value, and write numbers with increasing value to the number with the greatest value.

In order from least to greatest, these numbers read:

-8, -½, -0.02, 0.9, 2.5, 6.9

Better Grades & Higher Test Scores / MATH
©Incentive Publications, Inc., Nashville, TN

Get Sharp: Integers

Operations with Integers

A submarine dove
75 meters.
Then it descended
18 meters further.

-75 + -18 = -93 meters

Addition

The sum of two positive integers
is a positive integer.

$$1/10 + 6/10 = 7/10$$
$$6 + 14 = 20$$
$$0.4 + 1.7 = 2.1$$

The sum of 2 negative integers
is a negative integer.

$$-6 + -9 = -15$$
$$-8.3 + -4 = -12.3$$
$$-1/2 + -1/4 = -3/4$$

The sum of a positive integer and a negative
integer usually has the sign of the number
with the greater absolute value.

$$100 + -60 = 40$$
$$-38 + 10.5 = -27.5$$
$$1/2 + -7/8 = -3/8$$

I hope I can
remember what
I learned
about operating
on integers.

Subtraction

To subtract an integer, add its positive.

$$-6 - -3 = -6 + 3 = -3$$
$$10 - -4 = 10 + 4 = 14$$
$$-150 - -150 = -150 + 150 = 0$$
$$12 - -14 = 12 + 14 = 26$$
$$80 - 3 = 80 + -3 = 77$$

J.J. is $50 in debt to
his friend Sam.
Sam offers to remove
$15 of the debt.
What is J.J.'s balance?

-50 - -15 = -50 + 15 = -$35

Better Grades & Higher Test Scores / MATH
©Incentive Publications, Inc., Nashville, TN

Multiplication

The product of two positive integers
is a positive integer.

$7 \times 12 = 84$
$0.3 \times 6 = 1.8$
$\frac{1}{2} \times 3 = 1\frac{1}{2}$

The product of two negative integers
is a positive integer.

$-5 \times -0.5 = 2.5$
$-10 \times -20 = 200$
$-\frac{1}{4} \times -\frac{1}{2} = 1/8$

Zeke's overweight dog lost 3 pounds a week for 7 weeks. What was the total change in Fido's weight?

$-3 \times 7 = -21$ pounds

The product of a positive and
a negative integer
is a negative integer.

$200 \times -6 = -1200$
$-0.4 \times 16 = 64$
$6 \times -9 = -54$

Division

The quotient of two positive integers
is a positive integer.

$15 \div 5 = 3$
$670 \div 100 = 6.7$
$50 \div 0.5 = 100$

The quotient of two negative integers
is a positive integer.

$-12 \div -0.5 = 24$
$-4500 \div -9 = 500$
$-88 \div -11 = 8$

Angie's pet crocodile weighs 40 pounds less than he did 8 weeks ago. What was the average weight change per week?

$-40 \div 8 = -5$ pounds

The quotient of a positive
and a negative integer is
a negative integer.

$-560 \div 8 = -70$
$810 \div -90 = -8$
$-25.5 \div 0.5 = -51$

Get Sharp Tip # 30

Any integer divided by zero is zero.

A **mathematical expression**
is a phrase or statement
that uses symbols instead of words.

WORDS . . .
MATHEMATICAL EXPRESSIONS

three times a number **x** $3x$

•

eight less than a number **y** $y - 8$

•

a number **b** divided by negative 6 $b \div -6$

•

50 multiplied by a number **d** $50d$

•

20 divided by the sum of 7 and a number **k** $\dfrac{20}{7+k}$

•

negative 5 times the difference between a number **w** and 16 $-5(w-16)$

•

a number **n** decreased by 100 $n - 100$

•

12 times the product of 4 and a number **p** $12(4p)$

•

triple the sum of a number **q** and 10 $3(q+10)$

Terms are the numbers and variables in an expression.

5n – 7y + 3b has three terms: **5n**, **7y**, and **3b**.

A **variable** is a number represented by a letter in an expression.

In **12x**, the variable is **x**. In **145b**, the variable is **b**.

A **coefficient** is the number before the letter in an expression with a variable.

In **12x**, the coefficient is **12**. In **–70g**, the coefficient is **–70**.

Evaluating & Simplifying Expressions

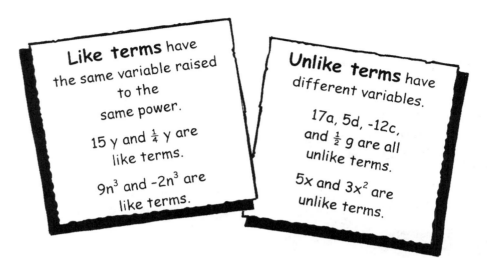

Like terms have the same variable raised to the same power.

15 y and $\frac{1}{4}$ y are like terms.

$9n^3$ and $-2n^3$ are like terms.

Unlike terms have different variables.

17a, 5d, -12c, and $\frac{1}{2}$ g are all unlike terms.

5x and $3x^2$ are unlike terms.

DAZED & CONFUSED

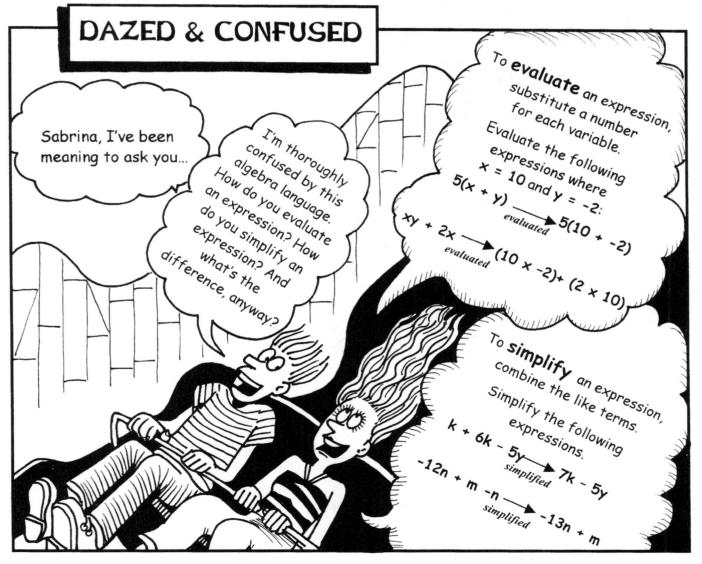

Sabrina, I've been meaning to ask you...

I'm thoroughly confused by this algebra language. How do you evaluate an expression? How do you simplify an expression? And what's the difference, anyway?

To **evaluate** an expression, substitute a number for each variable. Evaluate the following expressions where x = 10 and y = -2:

$5(x + y) \xrightarrow{evaluated} 5(10 + -2)$

$xy + 2x \xrightarrow{evaluated} (10 \times -2) + (2 \times 10)$

To **simplify** an expression, combine the like terms. Simplify the following expressions:

$k + 6k - 5y \xrightarrow{simplified} 7k - 5y$

$-12n + m - n \xrightarrow{simplified} -13n + m$

179

Equations

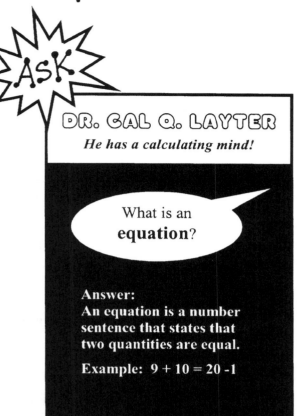

DR. CAL Q. LAYTER

He has a calculating mind!

What is an **equation**?

Answer:
An equation is a number sentence that states that two quantities are equal.

Example: $9 + 10 = 20 - 1$

Some equations include **variables** (quantities represented by a letter).

Example: $20n = 80$
                 n is the variable.

An equation is solved when you find the value of the variable. The value of the variable is called the **solution**.

Example: In $20n = 80$, $n = 4$
                 the solution to $20n = 80$ is 4.

Reading and Writing Equations

Tom (t) is 7 years older than Rob (r). $t = r + 7$

Sam's socks (s) have 3 more than twice
the number of holes as Abby's socks (a). $s = 2a + 3$

Michael (m) ran triple the amount of $m = 3b$
time as Britt (b).

Zeke (z) spent one-fourth the amount
of money as Chad (c) and Dana (d) spent together. $z = \frac{1}{4}(c + d)$

On Saturday (s), Cody rode his bike
6 miles farther than on Wednesday (w). $w + 6 = s$

Inverse Operations

An **inverse operation** is an operation that *reverses* or *undoes* another operation.

Add 5	*is the inverse of*	subtract 5.
Subtract 12	*is the inverse of*	add 12.
Multiply by 10	*is the inverse of*	divide by 10.
Divide by 28	*is the inverse of*	multiply by 28.

Get Sharp Tip #31
Learn inverse operations well. You will need them to solve equations!

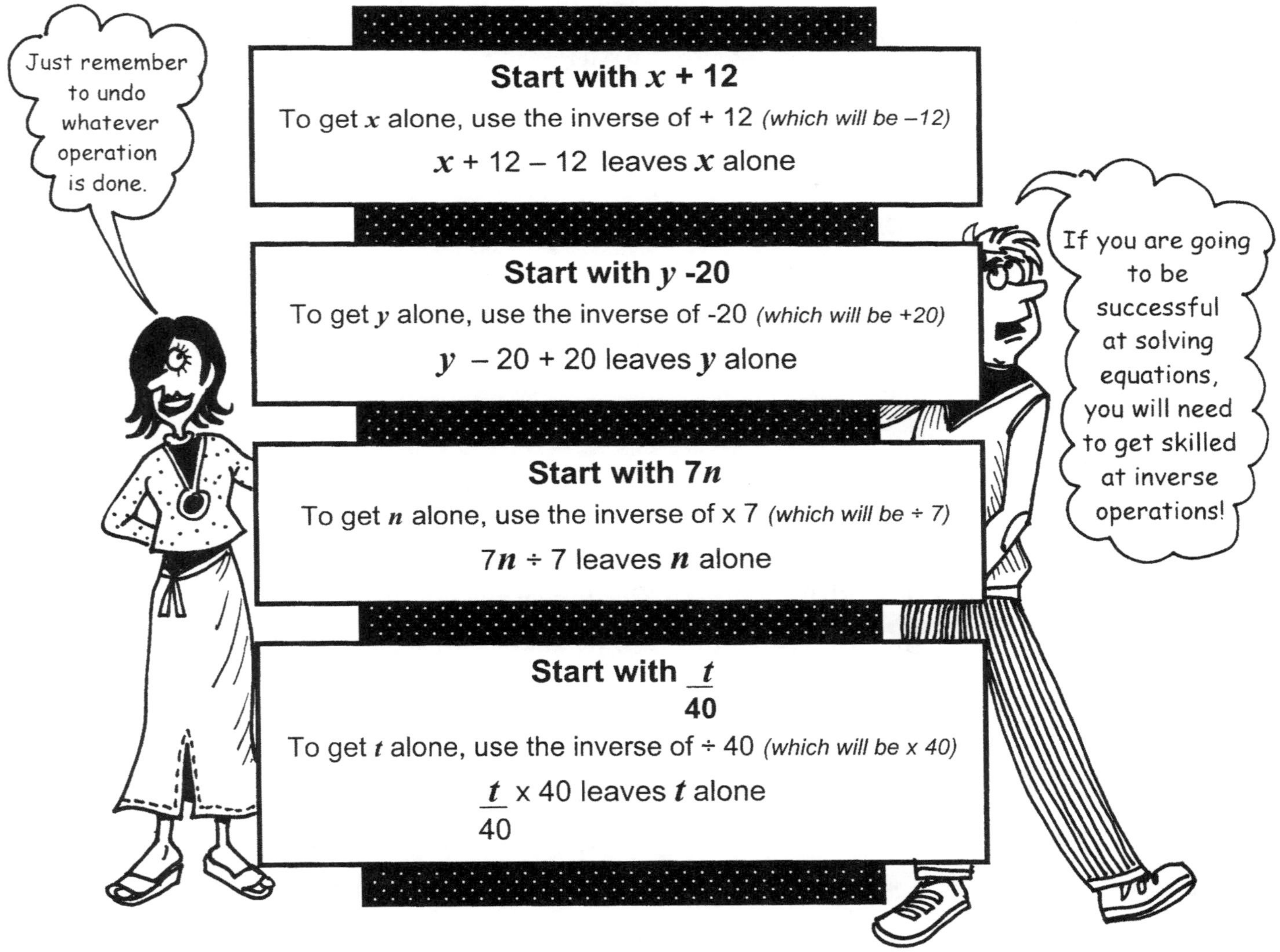

Just remember to undo whatever operation is done.

Start with $x + 12$

To get x alone, use the inverse of + 12 *(which will be −12)*

$x + 12 - 12$ leaves x alone

Start with y -20

To get y alone, use the inverse of -20 *(which will be +20)*

$y - 20 + 20$ leaves y alone

Start with $7n$

To get n alone, use the inverse of x 7 *(which will be ÷ 7)*

$7n ÷ 7$ leaves n alone

Start with $\dfrac{t}{40}$

To get t alone, use the inverse of ÷ 40 *(which will be x 40)*

$\dfrac{t}{40} \times 40$ leaves t alone

If you are going to be successful at solving equations, you will need to get skilled at inverse operations!

Solving Equations

To solve an equation, use inverse operations to get a simpler equation. This makes it easier to see a solution.

BE ALERT! BE AWARE!
Follow this rule carefully: When you use an inverse operation, ...

. . . . . you **MUST** do the same thing to **BOTH** sides of the equation!

HOW DO YOU SOLVE IT?

$d - 10 = 35$
$d - 10 + 10 = 35 + 10$ *Add 10 to both sides.*
$d = 45$ *Plus 10 cancels out minus 10, so **d** is alone.*

$n + \text{-}7 = 100$
$n + \text{-}7 - \text{-}7 = 100 - \text{-}7$ *Subtract −7 from both sides.*
$n = 107$ *The minus -7 cancels out plus -7, so **n** is alone.*

$20g = 380$
$20g \div 20 = 380 \div 20$ *Divide both sides by 20.*
$g = 19$ *Divide by 20 cancels out multiply by 20, so **g** is alone.*

$s \div 6 = 300$
$(s \div 6) \times 6 = 300 \times 6$ *Multiply both sides by 6.*
$s = 1800$ *Multiply by 6 cancels out divide by 20, so **s** is alone.*

Get Sharp: Expressions & Equations

Better Grades & Higher Test Scores / MATH
©Incentive Publications, Inc., Nashville, TN

Solving Multi-Step Equations

Some equations need more than one operation to find a solution.
Follow these steps to solve multi-step equations.

1) Simplify by combining any like elements.

2) Do the adding or subtracting first.

3) Do the multiplying or dividing last.

DON'T FORGET!

Your goal is to get the variable alone on one side of the equation.

$$10x + 6.5 = 106.5$$
$$10x + 6.5 - 6.5 = 106.5 - 6.5 \quad \text{Subtract 6.5 from both sides.}$$
$$10x = 100$$
$$10x \div 10 = 100 \div 10$$
$$x = 10 \quad \text{Divide both sides by 10.}$$

$$135 + f + 6f - 58 = 154$$
$$135 + 7f - 58 = 154 \quad \text{Combine f and 6f.}$$
$$77 + 7f = 154 \quad \text{Combine 135 and -58}$$
$$77 + 7f - 77 = 154 - 77 \quad \text{Subtract 77 from both sides.}$$
$$7f = 77$$
$$7f \div 7 = 77 \div 7 \quad \text{Divide both sides by 7.}$$
$$f = 11$$

$$30t = 5t + 150$$
$$30t - 5t = 5t - 5t + 150 \quad \text{Subtract 5t from both sides.}$$
$$25t = 150$$
$$25t \div 25 = 150 \div 25 \quad \text{Divide both sides by 25.}$$
$$t = 6$$

Solving Equations with Rational Numbers

DON'T FORGET!
Whatever you do to one side of the equation, you must do to the other side, too.

Equations with rational numbers are solved by the same rules as equations with whole numbers. (See pages 180–184.)

$$\frac{3}{5}b = 90$$

$$\frac{3}{5}b \div \frac{3}{5} = \frac{90}{1} \div \frac{3}{5}$$ *Divide both sides by $\frac{3}{5}$.*

$$\frac{3}{5}b \times \frac{5}{3} = \frac{90}{1} \times \frac{5}{3}$$ *To divide by a fraction, invert and multiply.*

$$b = 150$$

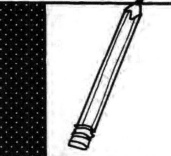

$$-1.9 + 60n - 10n = 198.1$$

$$-1.9 + 50n = 198.1$$ *Combine 60 n and –10 n.*

$$-1.9 + 50n - -1.9 = 198 - -1.9$$ *Subtract –1.9 from both sides.*

$$50n = 200$$ *Divide both sides by 50.*

$$n = 4$$

Three for you... and three for you...

$$2p - 7.6 = 22.4$$

$$2p - 7.6 + 7.6 = 22.4 + 7.6$$ *Add 7.6 to both sides.*

$$2p = 30$$

$$2p \div 2 = 30 \div 2$$ *Divide both sides by 2.*

$$p = 15$$

Problem-Solving with Equations

Equations come in very handy for solving all kinds of problems. If you're seeking a solution to a tricky word problem, try translating the problem into an equation. That will make the solution much easier to find. Here are two examples.

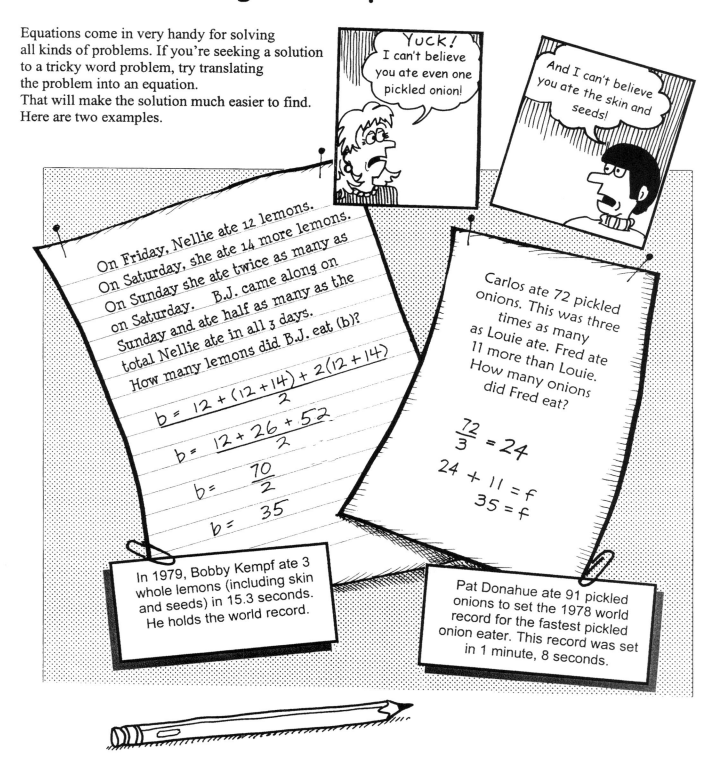

On Friday, Nellie ate 12 lemons. On Saturday, she ate 14 more lemons. On Sunday she ate twice as many as on Saturday. B.J. came along on Sunday and ate half as many as the total Nellie ate in all 3 days. How many lemons did B.J. eat (b)?

$$b = 12 + (12 + 14) + \frac{2(12 + 14)}{2}$$

$$b = \frac{12 + 26 + 52}{2}$$

$$b = \frac{70}{2}$$

$$b = 35$$

Carlos ate 72 pickled onions. This was three times as many as Louie ate. Fred ate 11 more than Louie. How many onions did Fred eat?

$$\frac{72}{3} = 24$$

$$24 + 11 = f$$

$$35 = f$$

In 1979, Bobby Kempf ate 3 whole lemons (including skin and seeds) in 15.3 seconds. He holds the world record.

Pat Donahue ate 91 pickled onions to set the 1978 world record for the fastest pickled onion eater. This record was set in 1 minute, 8 seconds.

Inequalities

Rex's boa constrictor is 11 feet long. My pet boa is shorter than Rex's. This inequality shows the length of my boa:

$$g < 11$$

An **inequality** is a number sentence that describes two quantities that are not equal.

There are many solutions to this inequality. Gigi's pet can be any length shorter than 11 feet. A number line can be used to show a graph of the possible solutions.

The open circle at 11 shows that 11 is not one of the possible solutions.

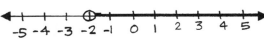

Charlie's boa is the size of Rex's or longer. This inequality shows the length of his boa:

$$c \geq 11$$

The closed circle at 11 shows that 11 is one of the possible solutions.

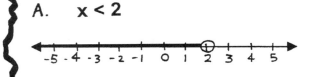

A. $x < 2$

C. $x > -2$

B. $x \geq -1$

D. $x \leq 3$

Better Grades & Higher Test Scores / MATH
©Incentive Publications, Inc., Nashville, TN

Get Sharp: Expressions & Equations

Equations with Two Variables

When in doubt. . . phone a friend

Roxie and I both have measles. Roxie (x) has 5 times as many as I do (y). We can write this equation about our measles:

$x = 5y$

This equation has two variables. How do we solve it?

Hello, Mom?

Sniff
Sniff

Well, dear, since neither quantity is known, you do have two variables. The value of x will depend on the value of y. For each value of x there is only one value of y. There will be several solutions. Each solution will be a pair of numbers (called a function). Make a table to find some of the solutions.

Or, take two aspirin and call me in the morning.

x = 5y

x	y	(x, y)
30	6	(30, 6)
40	8	(40, 8)
50	10	(50, 10)
65	13	(65, 13)
100	20	(100, 20)
500	100	(500, 100)

y = x + 3

x	y	(x, y)
-4	-1	(-4, -1)
-2	1	(-2, 1)
0	3	(0, 3)
5	8	(5, 8)
7	10	(7, 10)

x = -4y

x	y	(x, y)
4	-1	(4, -1)
8	-2	(8, -2)
-4	1	(-4, 1)
-12	3	(-12, 3)
-16	4	(-16, 4)

Get Sharp: Expressions & Equations

Better Grades & Higher Test Scores / MATH
©Incentive Publications, Inc., Nashville, TN

Graphing on a Coordinate Plane

A **coordinate plane** is formed by two lines, called **axes**, drawn perpendicular to each other for form a grid. An **ordered pair** of numbers (such as 5, 4) can be graphed as a point on a coordinate plane.

The pair of numbers gives the **coordinates** (location) of the point.

The **horizontal line** is the **x-axis**.

The **vertical line** is the **y-axis**.

The two axes meet at a point called the **origin**.

Get Sharp Tip # 32

When writing the coordinates, write the location on the x-axis first, followed by the location on the y-axis.

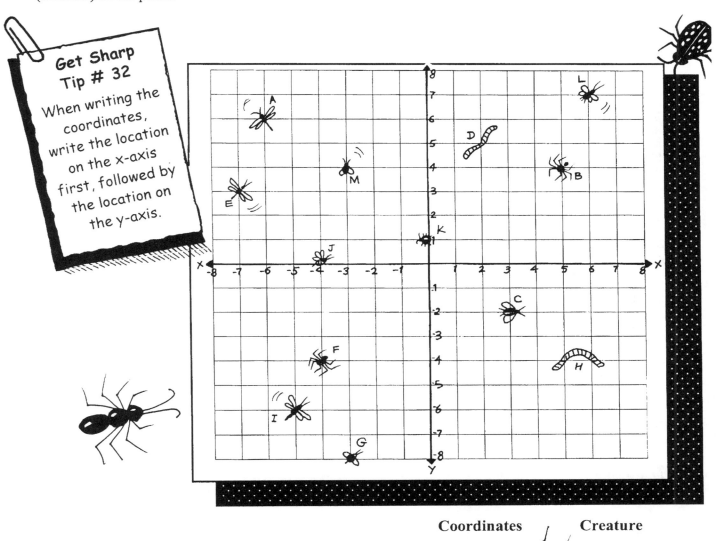

Coordinates	Creature
(-7, 3)	dragonfly (E)
(-4, 0)	bee (J)
(2, 5)	worm (D)
(3, -2)	fly (C)
(-4, -4)	spider (F)

Graphing Equations with 2 Variables

A graph of the equation **y = x + 2** consists of all the points on a coordinate plane whose coordinates are solutions to the equation.

Ordinarily, you cannot make a graph large enough to show all the solutions. You can make a graph that shows some of the solutions, and realize that the line would extend beyond your graph.

The table shows some of the solutions to **y = x + 2**. See how this equation looks when it is graphed.

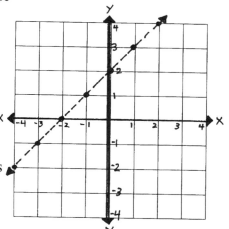

y = x + 2		
x	**y**	**(x, y)**
-4	-2	(-4, -2)
-3	-1	(-3, -1)
-2	0	(-2, 0)
-1	1	(-1, 1)
0	2	(0, 2)
1	3	(1, 3)
2	4	(2, 4)

An equation whose solutions form a straight line is called a **linear equation**.

Graph of y = 2x + 1

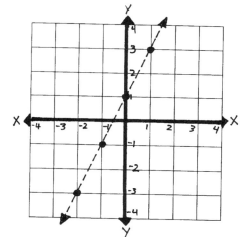

Graph of x = -y

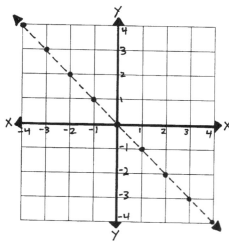

Better Grades & Higher Test Scores / MATH
©Incentive Publications, Inc., Nashville, TN

———GET SHARP→

on

Problem Solving

The problem is - this mouse hole is 2 inches wide and my head is 6 inches wide!

Approaching Problems

Defining the Problem

A problem is a question
to be answered.
When you read a problem,
try to identify
the problem clearly.
Ask yourself:
*What is the exact question
to be answered?
Can the problem be solved?*

People in the UK consume
more baked beans than in any
other country. They consume
11 lb, 11 oz per person,
per year. How many fewer
beans are consumed per
person in the US?

The question is:
*How many fewer beans are
consumed per person
in the US?*
There is not enough
information to solve
this problem.

In Switzerland, people
consume on the average of
1.5 kg per year of baked
beans. This is 3.6 kg less than
in Ireland. What is the average
consumption in Ireland?

The question is:
*What is the average baked
bean consumption in Ireland?*
This problem
can be solved.

Examining the Information

Identify the information needed to solve the problem.
Sometimes there is more than you need.
Sometimes information is missing.

There is not enough information.
To solve the problem, you need
to know how much is chewed
in Israel or Spain.

In Denmark, people chew more gum
than anywhere else in the world. They
consume twice as much gum as people
in Israel or Spain. On the average, how
much do the Danish chew each year?

There is too much information.
You do not need to know the baked
bean or meat consumption in order
to solve the problem.

New Zealand consumers love butter.
They each eat about 21 lbs a year. They
also consume over 4 lb of baked beans
and 242 lb of meat each. Butter lovers
in Switzerland consume two-thirds of
New Zealand's amount per person.
What is the Swiss average for butter
consumption per year?

Choosing the Operation

After the problem is identified and
carefully read, find out what operations are needed.

Watch for word clues that give a hint at the operation.

> The words *how many fewer*
> suggest that subtraction is the
> operation for this problem.

80,800,000 people in the US walked for exercise in 1999. 45,200,000 people reported that they exercised with equipment. How many fewer exercised using equipment?

> The words *five times* suggest that
> multiplication is the operation for
> this problem.

Florence Griffith Joyner was one of the fastest female runners ever. She ran 100 meters in 10.39 seconds. It takes me five times that long to run 100 meters. How long does it take me?

Choosing the Order of Operation

Many problems call for more than one operation. Sometimes it makes a difference which operation is done first. Look carefully at the problem before you decide the order of the operations.

A bat traveled 14 miles. An ostrich traveled 26 miles more. A sailfish swam twice as far as the bat and ostrich combined. A cheetah traveled a distance 30 miles less than the sailfish. How far did the cheetah travel?

First: *Add* 14 to 26 to find the ostrich's distance: $14 + 26 = 40$

Second: *Add* the bat's and ostrich's distances: $14 + 40 = 54$

Second: *Multiply* the sum by 2 to find the sailfish distance: $2 \times 54 = 108$

Third: *Subtract* 30 from the product: $108 - 30 = 78$ miles

Better Grades & Higher Test Scores / MATH
©Incentive Publications, Inc., Nashville, TN

Get Sharp: Approaching Problems

The Problem-Solving Process

The problem-solving process is an approach or series of steps that should be followed to solve a problem. Get familiar with this process, and practice it every time you have a problem to solve. It will pay off many times! You'll get sharper at finding accurate solutions to many kinds of problems. See the example on page 195 where the process was followed to solve a sample problem.

How to Solve a Math Problem

Step 1 **Read the problem carefully.**

Charlie has his own country western rock band, called **Charlie & the Rustlers**. They took a look at their record sales to find out how well they did last year. Here is some information about the sale of their albums. The total of all four albums sold last year was 284,000. What was the number of sales for the album, *Not Your Ordinary Cowboy Songs?*

Albums and Sales Amounts

Album 1:
Not Your Ordinary Cowboy Songs

Album 2:
I Lost My Head in Montana
sold 6000 fewer than "Not Your Ordinary Cowboy Songs"

Album 3:
Yes, Cowboys Can Sing
sold 3 times as many as "Not Your Ordinary Cowboy Songs"

Album 4:
You Broke My Heart on Friday
sold 20,000 more than "Not Your Ordinary Cowboy Songs"

Step 2 **Identify the problem to be solved or the question to be answered.**

How many copies of the album, *Not Your Ordinary Cowboy Songs*, sold last year?

Step 3 **Change the word problem into mathematical numbers, sentence, or idea.**

number sold of *Not Your Ordinary*. .	=	x
number sold of *I Lost My Head*. . .	=	$x - 6,000$
number sold of *Yes, Cowboys* . . .	=	$3x$
number sold of *You Broke My Heart*. . .	=	$x + 20,000$

Step 4   Choose a strategy that will work to solve the problem.

A strategy that should work well for this problem is to translate the word problem into a mathematical equation.
A diagram might also be helpful.

$$x + 3x + (x - 6{,}000) + (x + 20{,}000) = 284{,}000$$

Step 5   Use the strategy to find a solution.

$$x + 3x + (x - 6{,}000) + (x + 20{,}000) = 284{,}000$$
$$6x - 6{,}000 + 20{,}000 = 284{,}000$$
$$6x + 14{,}000 = 284{,}000$$
$$6x = 270{,}000$$
$$x = 45{,}000$$

Step 6   Communicate your process.

Be able to show (with number sentences, written calculations, drawings, or diagrams) how you reached the solution.

These are ways the process is communicated:
1) complete and careful equation and written work to solve it (shown in step 5)
2) expressions shown to represent each album (shown in step 3)
2) diagram with expressions shown represent each album (shown in step 4)

Step 7   Verify your answer.

Review the problem carefully. Make sure the answer is reasonable. Find a way to show that the answer is correct.

Check with addition.

Not Your Ordinary	=	45,000
I Lost My Head.	=	45,000 – 6,000
Yes, Cowboys	=	3 x 45,000
You Broke My Heart	=	45,000 + 20,000

195

Problem-Solving Strategies

A problem-solving strategy is a method for approaching and solving a problem. There are many different ways to solve problems. Different strategies fit well with different kinds of problems. One of the skills involved in sharp problem-solving is being able to choose a good strategy. Here are some strategies to have among your list of tools for attacking problems. (See pages 196-206).

—— Guess & Check

Sometimes the best strategy for solving a problem is to make a smart guess. After you make a careful guess, if it is possible, you can count or calculate to see if your guess was right.

The **Guess & Check** strategy is a good one for a problem where you can see a quantity, but it is too large, complex, or far away to count accurately and easily. Use it for this problem.

The Problem:

How many gumballs will the machine hold?

——Trial & Error

For some problems, the best strategy is to try out different solutions until you find one that works. **Trial & Error** is a good strategy for those tricky age problems.

The Problem:

Joe is 5 years younger than Moe.
2 years ago, Joe was 2/3 of Moe's age (at that time).
In 3 years, Joe will be ¾ of Moe's age (at that time).
In 20 years, the sum of their ages will be 69.
How old is Joe?

Boe tried several ages for Joe. After trying 5, 6, 7, 8, 9, 10, 11, and 12, he realized that the age 12 worked as a correct solution.

Better Grades & Higher Test Scores / MATH
©Incentive Publications, Inc., Nashville, TN

——Simplify

The many words and explanations in a problem might confuse you.
Some problems are like that! You can make a problem less complicated
by re-wording it into a shorter or simpler question.

The Problem:

Suki searched for a good deal on new
snowboards. She wanted to spend no more
than 20% of her savings and her next paycheck
combined. She has $499 in her savings
account. Her next paycheck is expected
to be $185. How much is Suki willing to
spend on a snowboard?

The problem simplified:

What is 20% of
($499 + $185)?

It's
$136.80.

——Find a Common Element

If the facts in a problem include different units, your first step is to convert all the
necessary information into the same unit. Often it is best to convert all facts to the
smallest common element.

The Problem:

Suki talked on the phone to one friend for 3 hours and 14 minutes
on Sunday, 73 minutes on Monday, and 39 minutes, 50 seconds
on Tuesday. She did have some short calls. On Wednesday and
Thursday, she talked to her friend just 40 seconds each day.
These calls were all long-distance. Her new phone card charges
2¢ for every 30 seconds. How much did Suki pay for the calls?

Since the fee for calls is rated in terms of seconds, it is best to convert all the times to seconds.
First, convert hours to minutes, then minutes to seconds.

Sun: 3 x 60 + 14 = 194 min x 60 = 11,640 seconds	Wed: 40 seconds
Mon: 73 x 60 = 4380 seconds	Thur: 40 seconds
Tue: 39 x 60 + 50 = 2340 + 50 = 2390 seconds	

That's only
$5.66!
What a deal!

Then, you can finish the calculations in the problem by adding all the times, dividing by 30,
and multiplying by 2¢

Better Grades & Higher Test Scores / MATH
©Incentive Publications, Inc., Nashville, TN

Get Sharp: Problem-Solving Strategies

—— Make a Model

Some questions and calculations are very difficult when you can't actually see the object in the problem. Creating a model is a handy problem-solving strategy for these situations. It is much easier to solve a problem and answer questions when you can actually see the characteristics of the object. The model might be built with paper, cardboard, toothpicks, straws, boxes, marshmallows, or many other common objects.

——— Draw a Diagram or Picture

Draw a diagram or a picture when you need to visualize a situation in order to solve a problem. This is much faster than making a model, and works in many situations where a model is too complicated or specific.

The Problem:

Five swimmers race for a buoy.
Nan is behind Dan but ahead of Van.
Van is between Stan and Jan.
Stan is behind Van.

Who is in the lead?

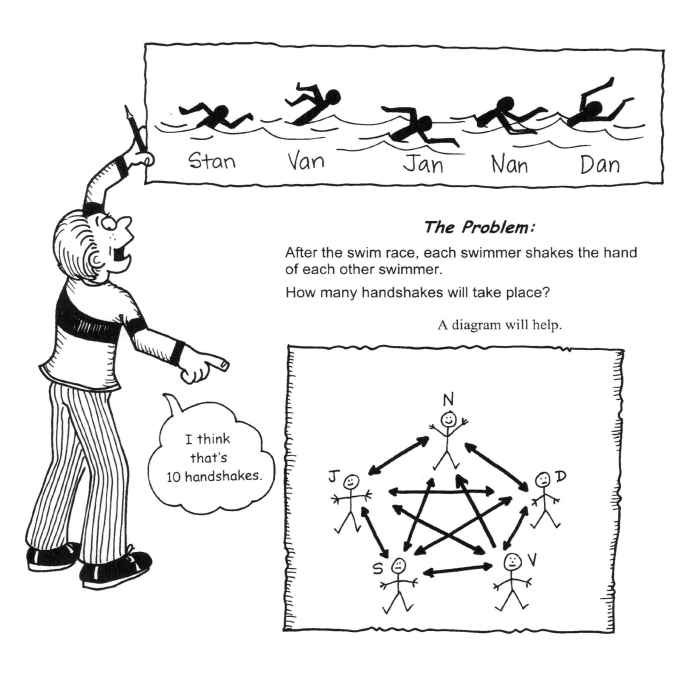

Stan   Van   Jan   Nan   Dan

I think that's 10 handshakes.

The Problem:

After the swim race, each swimmer shakes the hand of each other swimmer.

How many handshakes will take place?

A diagram will help.

— Make a Table or Graph

If there is a lot of data to consider in solving a problem, try putting it into a table. That way you can easily see relationships between numbers. A graph can also help you see relationships between numbers. A line graph is especially helpful if you need to see how amounts change over time.

The Problem:

Two ski racers, Sam and Pam, began practicing for the season by racing down the same hill six times a week. The first day of practice, Sam's race time was 3 minutes and 46 seconds. Pam's time was 4 minutes and 13 seconds. Each day of practice, Sam cut 2 seconds off his time, and Pam cut 3 seconds off her time.

Who was getting down the hill the fastest by the end of the 6th week?

Week	Sam	Pam
Start Time	3 min 46 sec	4 min 13 sec
Wk. 1	3 min 34 sec	3 min 55 sec
Wk. 2	3 min 22 sec	3 min 37 sec
Wk. 3	3 min 10 sec	3 min 19 sec
Wk. 4	2 min 58 sec	3 min 1 sec
Wk. 5	2 min 46 sec	2 min 43 sec
Wk. 6	2 min 34 sec	2 min 25 sec

The table can be used to write the data for racing times for each skier at the end of each week. Then, it only takes a glance to answer the question and solve the problem.

I think it's Pam!

The graph makes it easy to answer the question.

The Problem:

Sam and Pam also kept track of their hours of practice for 8 weeks.

Which skier had the steepest drop in hours of practice from one week to the next? Between which weeks did that drop occur?

Sam dropped off 6 hours between weeks 6 and 7.

— Estimate

An estimate is an approximate solution to a problem.
Sometimes, a problem does not need an exact answer,
and an estimation is a quick or practical solution.

The Problem:

The Grizzly's football team takes lot of showers during a season
Plenty of towels are needed for all those showers.
28 players shower each week after 6 practices and 1 game.
The season, including preseason practices, lasts 18 weeks.

Will 4500 clean towels be enough for the season?

> Round the 28 players to 30.
> Round the 6 practices + 1 game to 10.
> Round the 18 weeks of the season to 20.
> Multiply 30 x 10 x 20 to get 6000 towels.

Get Sharp
Tip # 33
Don't forget to use rounding when you estimate. It's a great estimation tool!

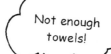

Not enough towels!

Extend a Pattern

Sometimes the best strategy is to look for a pattern in the data.
Then extend (continue) the pattern to find a solution to the problem.

The Problem:

Those long showers must be helping the team! They have had a great season, winning all their games so far. The table shows the scores in the first 7 games.

What do you predict the Grizzly's score will be in game 10?

Game	Opponent's Score	Grizzly's Score
1	0	3
2	3	9
3	6	15
4	9	21
5	12	27
6	15	33
7	18	39
8	21	45
9	24	51
10	27	57

The data has two patterns:

1) The opponent's score increased each time by 3 points.

2) In each game, the Grizzly's score was twice the opponent's plus 3.

Extend the pattern to find that the Grizzly's score in Game 10 will be a winning 57 points.

201

Work Backwards

Sometimes it is helpful to start at the end of a problem and work backwards to find a missing fact. This is especially useful when a problem has a missing fact somewhere in the middle (or at the beginning) or many items of data. Here is one problem that needs the backwards approach.

The Problem:

A busy rat walked 30 minutes into town to do errands. His first stop was the shoe store where he spent 12 minutes choosing new shoes and 21 minutes in line paying for them. Then it took 4 minutes to get to the library, where he read books for 1 hour. He walked 10 minutes to the gym, and did a workout that took 1 hour and 55 minutes. Then he jogged home in 18 minutes. He arrived home at 4:13 pm.

What time did he leave home?

I think I'll start at the end and work backwards on this one!

arrived home		4:13
left gym	−18 min	3:55
arrived at gym	− 1 hr, 55 min	2:00
left library for gym	−10 min	1:50
arrived at library	− 1 hr	12:50
left shoe store	− 4 min	12:46
got in line at shoe store	− 21 min	12:25
arrived at shoe store	− 12 min	12:13
left home for shoe store	− 30 min	11:43

Use a Formula

Some problems just need a formula. Usually a formula is a shortcut to a solution, so be alert for chances to use one. Make sure you choose the correct formula and use it accurately.

The Problem:

That hungry rat came home and ate a full box of cheese chips. Each cubic centimeter in the box holds 0.1 gram in weight of chips. What is the amount weight of the chips he ate?

Use the formula for volume of a rectangular prism.
$V = l \times w \times h$
$25 \times 10 \times 15 = 3750 \text{ cm}^3$

Then multiply the volume by 0.1 gram weight:
$3750 \times 0.1 = 375 \text{ grams}$

Use Mental Math

Your head is the best strategy for some math problems. You can think through a problem and come up with a solution without using any other tools. Here are some problems that can make good use of your mental gymnastics.

The Problems:

1. Lulu begins her daily workout at 6:55 am and finishes at 8:11 am. How long is her workout?

Start at 6:55 and mentally count the time until 8:11 am. Use your fingers if you need to.

> Exactly 1 hour and 16 minutes!

2. She stretches in 18 different poses, holding each one for 20 seconds. How many minutes does she spend stretching?

Multiply 18 x 2 in your head. Add a zero.

> That's easy— 360 seconds is 6 minutes!

3. Six times during her workout, Lulu stops for a drink of water. She drinks 195 mL each time. How many liters does she drink during a workout?

Multiply 6 x 200.
Subtract 6 x 5 (30, since 195 is 5 less than 200)
Convert 1170 mL to 11.7 L.

> She drank 1170 mL or 1.17 liters.

4. This week, Lulu did 120 sit-ups total. Last week, she did two-thirds this many. How many sit-ups did Lulu do last week?

Think of what 2/3 of 12 would be. (8) Add a zero.

> That's 80 sit-ups last week!

⎯ Write an Equation

One of the handiest problem-solving strategies is translating a problem into an equation. When the problem has a mixture of numbers and words, and especially when there is an unknown quantity, use an equation to solve the problem.

The Problem:

Twenty-six friends joined a slug-racing team.
Two dropped out each week for the first eight weeks.
Then five new members joined, giving the team
10 more members than last year.

How many members were on the team last year?

Write an equation carefully to make sure it is the right one to find a solution. One of these equations is a good one for this problem. Which one is it?

A. $26 + 8 + 2 + 5 + 10 = x$
B. $26 - (8 \times 2) + 5 = x + 10$
C. $x = 26 - (8 \times 2) + 5 + 10$

⎯ Write a Proportion

If a problem gives a ratio and asks for a solution that is at the same rate, set up a proportion to help find the solution.

The Problem:

20 slugs raced in the first group.
6 of these got injured during the race.
A total of 80 slugs will race during
the 2-day competition.

At this injury rate, how many slugs will get injured during the races?

Set up this proportion to solve the problem:

$$\frac{6}{20} = \frac{x}{80}$$

$$20x = 480$$

$$x = 24$$

Use Logic

Logic problems make use of reasoning skills. They use *If....Then* thinking.
When a problem requires you to use some information
to make assumptions about other information, you need to use logic.

Follow along with the steps in reasoning taken to solve this problem.

The Problem:

Each of the four hikers
has one of these
ailments: blisters,
shoulder bruises,
a swollen ankle,
or bee stings.

Read the clues,
then find the answer
to the question:

The Clues:

The hiker with the bee stings is wearing a hat.

The hiker with the blisters has no hat.

The hiker with bruised shoulders has no sunglasses.

The hiker with a swollen ankle is not wearing long pants.

The hiker with bruised shoulders has no hat.

Reasoning steps by one problem-solver: *(There are other ways to reason this problem.)*

1) Hiker with bee stings must be B or C (They have hats.)
2) Hiker with blisters must be A or D (They have no hats.)
3) Hiker with bruised shoulders cannot be A, B, or C. (A has sunglasses, B and C have hats.)
4) Hiker with swollen ankle must be A, B, or D (They have no long pants.)

Therefore. . .   *Hiker with bruised shoulders is D.*
Hiker A has blisters, because D is taken.
Hiker B must have swollen ankle, because A and D are taken.

Solving Open-Ended Problems

Open-ended problems are mathematical problems that have more than one solution. These problems are great fun to solve. It's a challenge to try to find many solutions.
Here are a few examples:

The Problem:

Three friends spent a lot of time catching interesting creatures, such as insects and spiders. Wes caught 1 more than ½ as many as Bess, who caught 3 times as many as Les. Les caught a number that has 2 different even digits and is > 20 but < 30. How many creatures did Wes catch?

Find at least two different answers.

> Wes's number could be 37 **OR** 40.

The Problem:

Bess and Less went to buy some jars for their creatures. They each had a total of 10 coins. They each had a total of 76¢. They did not have the exact same 10 coins. What 10 coins could make 76¢?

Find at least two different answers.

> Two possibilities for 10 coins:
> 2 quarters, 2 dimes, and 6 pennies
> **OR**
> 6 dimes, 3 nickels, and 1 penny

The Problem:

On the way home, the friends ran out of gas, and had to walk to find some. They brought back 4 cans of gas with a total of 16 gallons. All the cans held a specific number of gallons (no fractions). All the cans were of different sizes. What sizes were the cans?

Find at least two different answers.

> Containers could be:
> a 2-gal,
> a 3-gal,
> a 5-gal,
> a 6-gal, ..

> ... OR,
> a 1-gal,
> a 4 gal,
> a 5-gal,
> a 6-gal.

Reviewing Your Answers

Check for Reasonableness: *Is the answer reasonable? Does it make sense?*

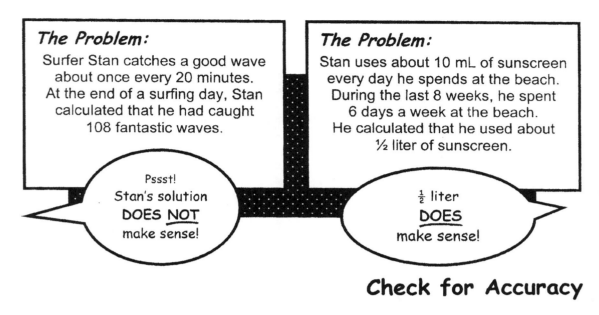

The Problem:
Surfer Stan catches a good wave about once every 20 minutes. At the end of a surfing day, Stan calculated that he had caught 108 fantastic waves.

Pssst! Stan's solution DOES <u>NOT</u> make sense!

The Problem:
Stan uses about 10 mL of sunscreen every day he spends at the beach. During the last 8 weeks, he spent 6 days a week at the beach. He calculated that he used about ½ liter of sunscreen.

½ liter DOES make sense!

Check for Accuracy

It's important to check every problem solution to see if it is correct. You can do this in many ways. Here are a few: check it with an inverse operation, work the problem backwards, or use a different strategy to solve it.

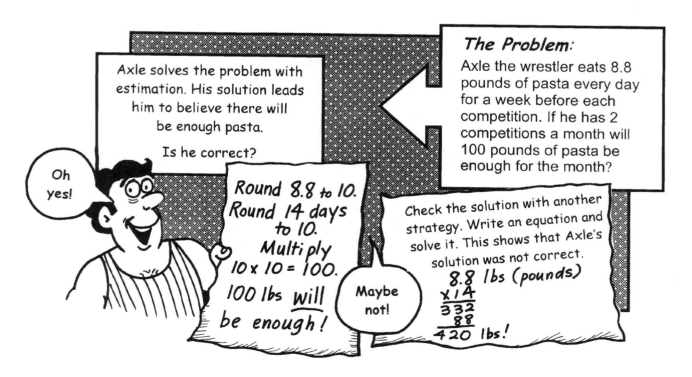

Axle solves the problem with estimation. His solution leads him to believe there will be enough pasta.

Is he correct?

The Problem:
Axle the wrestler eats 8.8 pounds of pasta every day for a week before each competition. If he has 2 competitions a month will 100 pounds of pasta be enough for the month?

Oh yes!

Round 8.8 to 10. Round 14 days to 10. Multiply 10 x 10 = 100. 100 lbs <u>will</u> be enough!

Maybe not!

Check the solution with another strategy. Write an equation and solve it. This shows that Axle's solution was not correct.

$$\begin{array}{r} 8.8 \text{ lbs (pounds)} \\ \times 14 \\ \hline 332 \\ 88 \\ \hline 420 \text{ lbs!} \end{array}$$

Better Grades & Higher Test Scores / MATH
©Incentive Publications, Inc., Nashville, TN

Get Sharp: Reviewing Problems

Problem-Solving Scoring Guide

PROBLEM SOLVING PROCESS SCORING GUIDE

TRAIT	SCORE OF 5	SCORE OF 3	SCORE OF 1
CONCEPTUAL UNDERSTANDING	• Student's work shows that the problem is clearly identified and understood. • Work clearly shows that the student has translated the written problem-solving task effectively into mathematical ideas.	• Student's work shows that the problem is identified and understood. • The student has done an adequate job of translating the written problem-solving task into mathematical ideas.	• Student's work does not show a clear identification or understanding of the problem. • The student has done a partial or incorrect job of translating the written problem-solving task into mathematical ideas.
STRATEGIES & PROCESSES	• Student has chosen appropriate strategies for solving the problem. • The strategies have been used in a complete, clear, and complex manner to move toward a problem solution. • Equations, symbols, models, pictures, and/or diagrams are clear and complete.	• Student has chosen appropriate strategies for solving the problem. • The strategies have been used in a complete, clear, and complex manner to move toward a problem solution. • Equations, symbols, models, pictures, and/or diagrams are complete and relatively clear.	• Student has not chosen appropriate strategies for solving the problem or has chosen appropriate strategies but not used them correctly or effectively. • Equations, symbols, models, pictures and/or diagrams are incomplete or do not lead to the solution.
COMMUNICATION	• The student has used words, symbols, pictures, models, and/or other graphics to clearly show the steps to a solution of the problem. • The student's explanation of the use of strategies and of the path taken to solution is clear and sensible.	• The student has used words, symbols, pictures, models, and/or other graphics to adequately show the steps to a solution of the problem. • The student's explanation of the use of strategies and of the path taken to solution is adequate.	• The student has not adequately used words, symbols, pictures, models, and/or other graphics to clearly show the steps to a solution of the problem. • The communication of the student's processes is skimpy, or nonexistent.
CORRECTNESS (Accuracy of the Answer)	• The student's answer is correct. • The student's work supports the answer given.	• The student's answer is mostly correct, with only minor errors. • The student's work supports the answer.	• The student's answer is incomplete, or incorrect. *and/or* • The student's work does not support the answer given.
VERIFICATION	• The student's work shows that he/she has reviewed the problem-solving process and made a clear, effective attempt to justify the answer or arrive at it in a different way. • The review supports the student's solution.	• The student's work shows that he/she has reviewed the problem-solving process and made an attempt to justify the answer. • The review supports the student's solution.	• The student's work does not show an effective or complete review of his/her process, or a defense or support of his/her solution.

A score of 4 may be given for papers that fall between 3 and 5 on a trait. A score of 2 may be given for papers that fall between 1 and 3.

Better Grades & Higher Test Scores / MATH
©Incentive Publications, Inc., Nashville, TN

—— GET SHARP →

on

Math Terms

Absolute Value — the distance a number is from 0 on the number line

Abundant Number — any number for which the sum of its factors (other than the number itself) is greater than itself

Additive Inverse — for a given number, the number that can be added to give a sum of 0. –4 is the additive inverse of +4 because –4 + (+4) = 0

Base — a standard grouping of a numeration system
(If a numeration system groups objects by fives, it is called a base 5 system; in a base 5 system, the numeral 23 means two fives and three ones.)

Cardinal Number — the number of elements in a set

Composite Number — a number having at least one whole number factor other than 1 and itself

Coefficient — the number amount in a mathematical expression
In the expression **5x**, **5** is the coefficient of **x**.

Digit — a symbol used to write numerals (In the decimal system, there are ten digits: 0, 1, 2, 3, 4, 5, 6, 7, 8, 9.)

Disjoint Sets — sets having no members in common
The sets {1, 9, 7, 12} and {3, 8, 11, 22} are disjoint sets.

Elements — the members of a set

Empty Set — a set having no elements, also called a null set
{ } or Ø represents an empty set.

Equivalent Sets — sets having the same number of members

Even Number — one of the set of whole numbers having 2 as a factor

Expanded Notation — the method of writing a numeral to show the value of each digit
5327 = 5000 + 300 + 20 + 7

Exponent — a numeral telling how many times a number is to be used as a factor
In 6^3, the exponent is 3
$6^3 = 6 \times 6 \times 6 = 216$

Finite Set — a set having a specific number of elements
{2, 5, 9, 15} is a finite set.

Inequality — a number sentence showing that two groups of numbers stand for different numbers
The signs ≠, <, and > show *inequality*.
7 + 5 > 12 – 9

Infinite Set — a set having an unlimited number of members

Integer — any member of the set of positive or negative counting numbers and O
(. . . –4, –3, –2, –1, O, 1, 2, 3, 4, . . .)

Irrational Number — a decimal that neither terminates nor repeats
Pi (π) and most square roots are examples of *irrational* numbers.

Intersection of Sets — the set of members common to each of two or more sets

The intersection of these sets is 3, 7, and 8.

  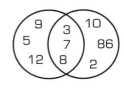 The symbol ∩ represents intersection.

Mixed Numeral — a numeral that includes a whole number and a fractional number, or a whole number and a decimal number

$7\frac{1}{2}$ and 37.016 are *mixed numerals*

Multiplicative Inverse — for any given number, the number that will yield a product of 1

$\frac{4}{3}$ is the *multiplicative inverse* of $\frac{3}{2}$ because $\frac{4}{3} \times \frac{3}{4} = 1$

Negative Integer — one of a set of counting numbers that is less than O

Number — a mathematical idea concerning the amount contained in a set

Number Line — a line that has numbers corresponding to points along it

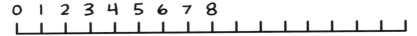

Numeral — a symbol used to represent or name a number

Numeration System — a system of symbols used to express numbers

Numerator — the number above the line in a fraction

Odd Number — a whole number belonging to the set of numbers equal to (n x 2) + 1
(Odd numbers are not divisible by 2.)

(1, 3, 5, 7, 9 . . .) are *odd numbers*.

Opposite Property — a property that states that if the sum of two numbers is 0, then each number is the opposite of the other
−4 + 4 = 0; so −4 and 4 are opposites

Ordered Pair — a pair of numbers in a certain order, with the order being of significance

Ordinal Number — a number telling the place of an item in an ordered set (sixth, eighth, etc.)

Origin — the beginning point on a number line (the origin is often 0)

Palindrome — a number which reads the same forward and backward (121, 343, 5995, 87678, 91219, etc.)

Periods — groups of three digits in numbers

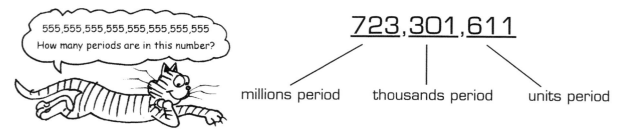

555,555,555,555,555,555,555,555
How many periods are in this number?

$$723{,}301{,}611$$

millions period   thousands period   units period

Place Value — the value assigned to a digit due to its position in a numeral

Positive Integer — one of a set of counting numbers that is greater than 0

Powers of a Number — the number of times a number is multiplied by itself
This is indicated by a small, superscript number next to the main number.
10^4 means *10 x 10 x 10 x 10*

Property of One — a property that states that any number multiplied by 1 will equal that number

Property of Zero — a property that states that any number plus zero equals that number

Prime Factor — a factor that is a prime number
(1, 2, and 5 are *prime factors* of 20.)

Prime Number — a number whose only number factors are 1 and itself

Rational Numbers — a number that can be written as the quotient of two numbers
(A terminating or repeating decimal is rational.)

Real Numbers — any number that is a positive number, a negative number, or 0

Reciprocals — a pair of numbers whose product is one
($\frac{1}{2}$ and $\frac{2}{1}$ are reciprocals.)

Reciprocal Method — (for dividing fractions) a means of dividing fractions that involves replacing the divisor with its reciprocal, then multiplying

$$\frac{2}{3} \div \frac{4}{7} = \frac{2}{3} \times \frac{7}{4} = \frac{14}{12} = \frac{7}{6} = 1\frac{1}{6}$$

Rename — to name numbers with a different set of numerals
(592 can be renamed as 500 + 90 + 2.)

Roman Numerals — numerals used by the Romans for keeping records

Rounding — disregarding all digits in a number beyond a certain place value

Scientific Notation — a number expressed as a decimal number (usually with an absolute value less than 10) multiplied by a power of 10

$$4.35 \times 10^3 = 4350$$

Sequence — a continuous series of numbers ordered according to a pattern

Set — a collection of items called members or elements

Skip Count — counting by skipping a certain number of digits
(counting by 2's, 5's, and 10's, etc.)

Subset — every member of a set, or any combination of the members of a set

Union of Sets — a set containing the combined members of two or more sets

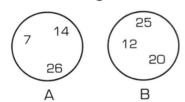

(The union of sets A and B is 7, 12, 14, 20, 25, 26.)

The symbol ∪ represents union.

Unit — the first whole number

Venn Diagram — a pictorial means of representing sets and the union or intersection of sets

  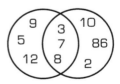

Whole Number — a member of the set of numbers (0, 1, 2, 3, 4 . . .)

Zero — the number between the set of all positive numbers and the set of all negative numbers

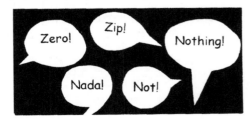

Get Sharp: Math Terms

Addend — a number being added in an addition problem
In the equation **4 + 7 = 11**, 4 and 7 are addends

Addition — an operation combining two or more numbers

Algorithm — method commonly used for performing computations involving mathematical operations (any computational procedure such as addition or multiplication)

Associative Property — (for addition and multiplication) the rule stating that the grouping of addends or factors does not affect the sum or product

$$(3 + 6) + 9 = 3 + (6 + 9)$$

$$(2 \times 4) \times 7 = 2 \times (4 \times 7)$$

Average — the sum of a set of numbers divided by the number of addends
The average of   1, 2, 7, 3, 8, and 9 =
$$\frac{1 + 2 + 7 + 3 + 8 + 9}{6} = 5$$

Binary Operation — any operation involving two numbers

Common Factor — a whole number that is a factor of two or more numbers
(3 is a factor common to 6, 9, and 12.)

Common Multiple — a whole number that is a multiple of two or more numbers
(12 is a multiple common to 2, 3, 4, and 6.)

Commutative Property — (for Addition and Multiplication) the rule stating that the order of addends or factors has no effect on the sum or product
3 + 9 = 9 + 3 and 4 x 7 = 7 x 4

Diagram — a 2-dimensional representation of an item or situation
Diagrams are often used to help with problem solving.

Difference — 1. the distance between two numbers on the number line;
2. the result of subtracting a lesser number from a greater number

In the equation **99 – 46 = 53**, 53 is the difference.

Distributive Property for Multiplication over Addition — the rule stating that when the sum of two or more addends is multiplied by another number, each addend must be multiplied separately and then the products must be added together
3 x (4 + 6 + 9) = (3 x 4) + (3 x 6) + (3 x 9) = 12 + 18 + 27

Dividend — a number that is to be divided in a division problem
In the equation **63 ÷ 7 = 9**, 63 is the dividend.

Divisibility — a number is divisible by a given number if the quotient of the two numbers is a whole number
(189 is divisible by 9 because 189 ÷ 9 is a whole number)

Division — the operation of finding a missing factor when the product and one factor are known

Divisor — the factor used in a division problem for the purpose of finding the missing factor
In this problem, the divisor is 12.

$$12\overline{)24}\,^{2}$$

Equation — a mathematical sentence which states that two expressions are equal
7 x 9 = 3 + (4 x 15)

Estimate — an approximation or rough calculation

Factor — one of two or more numbers that can be multiplied to find a product
In the equation **6 x 9 = 54**, 6 and 9 are factors.

Factor Tree — a pictorial means of showing the factors of a number

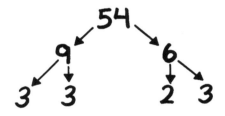

I didn't know that numbers grew on trees!

Greatest Common Factor — the largest number that is a factor of two other numbers
6 is the greatest common factor of 18 and 24.

Identity Element for Addition — 0 is the identity element for addition because 0 plus any given number equals that number.
3 + 0 = 3

Identity Element for Multiplication — 1 is the identity element for multiplication because 1 multiplied by any given number equals that number.
17 x 1 = 17

Least Common Multiple — the smallest whole number that is divisible by each of two or more given numbers
(The **least common multiple** of 2, 6, 9, and 18 is 18.)

Logic — "If–then" reasoning often used to solve math problems.
When using logical thinking, the problem solver thinks this way:
If x is true, then y must also be true.

Models — 3-dimensional replications of a figure or setting
Models are often used to help with problem solving.

Multiple — the product of two whole numbers

Multiplication — an operation involving repeated addition
5 x 4 = 4 + 4 + 4 + 4 + 4

Multiplicative Inverse — for any given number, the number that will yield a product of 1
$\frac{4}{3}$ **is the multiplicative inverse of** $\frac{3}{4}$ **because** $\frac{4}{3}$ **x** $\frac{3}{4}$ **= 1.**

Multi-step Problems — problems which take more than one step or more than one operation to solve

Open-Ended Problems — problems which have more than one correct answer
This is an open-ended problem:
What even, 3-digit numeral has digits whose sum is 12?

Prime Factor — a factor that is a prime number
1, 2, and 5 are prime factors of 20.

Product — the answer in a multiplication problem
In this problem, the product is 20,000.
4 x 5000 = 20,000

Quotient — the answer in a division problem
In this problem, the quotient is 75.
15,075 ÷ 201 = 75

Rename — to name numbers with a different set of numerals

Remainder — the number (less than the divisor) that is left
after a division problem is completed
In this problem, the remainder is 6.

$$\begin{array}{r} 20 \\ 21\overline{)426} \\ -42 \\ \hline 6 \end{array}$$

Subtraction — the operation of finding a missing addend when one addend and the sum are known

Strategy — a method used to approach and solve a problem

Sum — the answer in an addition problem resulting from the combination of two or more addends

428,108,970,492,419
+253,795,094,246,355
681,904,064,738,774

Now, that's some sum!

Better Grades & Higher Test Scores / MATH
©Incentive Publications, Inc., Nashville, TN

Good Words to Know for Fractions & Decimals

Let's see . . . ½ a mouse = 75% of 1 dinner.

Common Denominator — a whole number that is the denominator for both members of a pair of fractions

For $\frac{3}{7}$ and $\frac{5}{7}$, 7 is a common denominator.

Common Factor — a whole number which is a factor of two or more numbers

3 is a factor common to 6, 9, and 12.

Common Multiple — a whole number that is a multiple of two or more numbers

12 is a multiple common to 2, 3, 4, and 6.

Complex Fraction — a fraction having a fraction or a mixed numeral as its numerator and/or denominator
$$\frac{\frac{1}{5}}{\frac{1}{3}}$$

Cross Product Method (or Cross Multiplication) — a means of testing for equivalent fractions

if $\frac{3}{5} = \frac{6}{10}$, then 3 x 10 will equal 5 x 6.

Decimal Numeral — a name for a fractional number expressed with a decimal point, such as .27

4.03 is a mixed decimal

Denominator — the bottom number in a fraction; the denominator tells how many parts there are in a whole unit

Equivalent Fractions — fractions that name the same fractional number

$\frac{3}{4}$ and $\frac{9}{12}$ are equivalent.

Fraction — the name for a fractional number written in the form $\frac{a}{b}$; a is the numerator, b is the denominator

Fractional Number — a number that can be named as a fraction, $\frac{a}{b}$. The numerator and denominator can be any numbers with the exception that the denominator cannot be 0 .

Greatest Common Factor — the largest number that is a factor of two other numbers

6 is the greatest common factor of 18 and 24.

Improper Fraction — a fraction having a numerator equal to or greater than the denominator, therefore naming a number of 1 or more

$\frac{9}{4}$ is an improper fraction.

Least Common Denominator — the smallest whole number that is a multiple of the denominators of two or more fractions

The least common denominator for $\frac{1}{3}$ and $\frac{3}{4}$ is 12.

Least Common Multiple — the smallest whole number that is divisible by each of two or more given numbers

The least common multiple of 2, 6, 9, and 18 is 18.

Like Fractions — fractions having the same denominator

$\frac{2}{9}$ and $\frac{12}{9}$ are like fractions.

Lowest Terms — When a fraction has a numerator and denominator with no common factor greater than 1, the fraction is in lowest terms.

$\frac{3}{7}$ is a fraction in lowest terms.

Mixed Numeral — a numeral that includes a whole number and a fractional number, or a whole number and a decimal

$7\frac{1}{2}$ and 37.016 are mixed numerals.

Numerator — the number above the line in a fraction

Percent — a comparison of a number with 100, expressed using the % symbol

45% is a comparison of 45 to 100.

Proportion — a number statement of equality between two ratios

$$\frac{3}{7} = \frac{9}{21}$$

Rate — a comparison of two quantities

Ratio — a comparison of two numbers expressed as $\frac{a}{b}$

Reciprocals — two numbers whose product is one

$\frac{1}{3}$ and $\frac{3}{1}$ are reciprocals because $\frac{1}{3} \times \frac{3}{1} = 1$.

Reduced Fraction — a fraction whose numerator and denominator are not divisible by any number other than 1 $\quad \frac{3}{7}$

Repeating Decimal — a decimal in which a certain set of digits repeats without end

(0.363636)

Terminating Decimal — a decimal that shows a quotient of a whole number and a power of 10

$0.0204 = \frac{204}{1000}$ $\qquad$ $3.56 = \frac{356}{100}$

Terms of a Fraction — the numerator and denominator of a fraction

Better Grades & Higher Test Scores / MATH
©Incentive Publications, Inc., Nashville, TN

What formula will measure the capacity of my empty tummy?

Adjacent Angles — angles that have the same vertex and a common side between them

Angle A is **adjacent** to angle B.

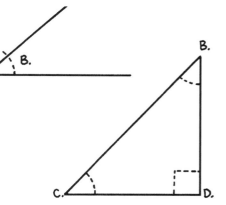

Adjacent Side — the leg next to a given angle in a right triangle

Side $\overline{CD}$ is **adjacent** to angle C.

Altitude of a Triangle — the distance between a point on the base and the vertex of the opposite angle, measured along a line that is perpendicular to the base (The altitude is also referred to as the height of the triangle.)

Segment $\overline{XY}$ is the **altitude** in this triangle.

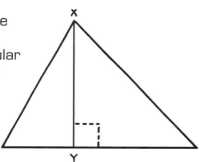

Angle — a figure formed by two rays having a common endpoint (vertex)

An **acute** angle — measures less than 90° (figure #1).

A **right** angle — measures 90° (figure #2).

An **obtuse** angle — measures more than 90° and less than 180° (figure #3).

A **straight** angle — measures 180° (figure #4).

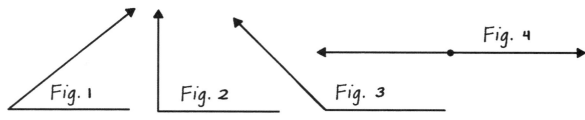

Fig. 1 Fig. 2 Fig. 3 Fig. 4

Central Angle — an angle formed by two radii of a circle.

Angle M is a **central angle**.

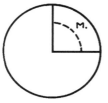

Complementary Angles — two angles whose combined measurements equal 90°
X and Y are **complementary** angles

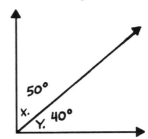

50°
X.
Y. 40°

Congruent Angles — angles having the same measure

Corresponding Angles — angles that are formed when a line intersects two parallel lines; corresponding angles are congruent. B and F (below) are **corresponding** angles

Supplementary Angles — two angles whose combined measurements equal 180° A and B (below) are **supplementary** angles.

Vertical Angles — angles that are formed opposite one another when two lines intersect; vertical angles are congruent. E and H (below) are **vertical** angles

That's a new angle!

A. B.
C. D.
E. F.
G. H.

Arc — a part of a circle between any two points on the circle

Segment $\overline{QR}$ is an **arc**.

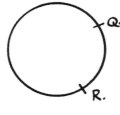

Q.

R.

Area — the measure of the region inside a closed plane figure (area is measured in square units)

Axis — a number line which may be vertical or horizontal

Axes — two perpendicular number lines with a common origin

Base — a side of a geometric figure

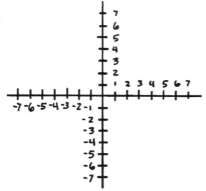

220

Bisect — to divide into two congruent parts

Bisector — a line or ray that divides a segment or angle into two congruent parts

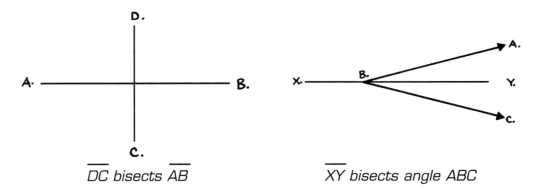

$\overline{DC}$ bisects $\overline{AB}$ $\overline{XY}$ bisects angle ABC

Capacity — the measure of the amount that a container will hold

Chord — a line segment having endpoints on a circle

$\overline{XY}$ is a chord.

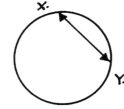

Circle — a closed curve in which all points on the edge are equidistant from a given point in the same plane (See circle above.)

Circumference — the distance around the outside edge of a circle
The measure of circumference = π x diameter

Closed figure — a set of points that encloses a region in the same plane; a curve that begins and ends at the same point

   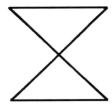

Coincide — Two lines coincide when they intersect at more than one point.

Collinear — When points are on the same line, they are collinear.

Compass — a tool for drawing circles

Cone — a space figure with a circular base and a vertex

Congruent — of equal size

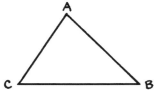 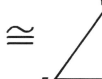 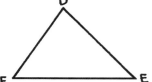

The symbol ≅ means *congruent*

Triangles ABC and DEF are **congruent**.

Coplanar — lines or points that are not in the same plane

Cube — a space figure having six congruent, square faces

Curve — a set of points connected by a line segment

Customary Units — units of the measurement system commonly used in a given country
Inches, feet, yards, miles, ounces, pounds, pints, quarts, and gallons are customary units in the U.S.

Cylinder — a space figure having two congruent, circular bases

Decagon — a ten-sided polygon

Degree — 1. a unit of measure used in measuring angles
(a circle is 360 degrees)
2. a unit for measuring temperature

Diagonal — a line segment joining two nonadjacent vertices in a polygon

$\overline{AC}$ is a **diagonal** in this figure.

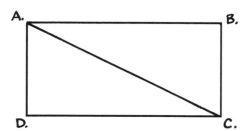

Diameter — a line segment which has its endpoints on a circle and which passes through the center of the circle

$\overline{LM}$ is the **diameter** of this circle.

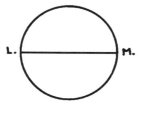

Dodecahedron — a space figure with 12 pentagonal faces

Edge — a line segment formed by the intersection of two faces of a geometric space figure

Endpoint — a point at the end of a line segment or ray

G is the **endpoint** of this ray.

Equilateral — having sides of the same length

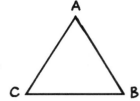

Figure ABC is an **equilateral** triangle.
All of its sides are the same length.

Face — a plane region serving as a side of a space figure

Flip — to "turn over" a geometric figure

The size or shape of the figure
does not change.

Geometry — the study of space and figures in space

Gram — a standard unit for measuring mass in the metric system

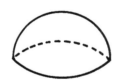

Hemisphere — half of a sphere

Heptagon — a seven-sided polygon

Hexagon — a six-sided polygon

Horizontal — a line that runs parallel to a base

Line $\overline{GH}$ is a **horizontal** line.

Hypotenuse — the longest side of a triangle located opposite the right angle

Side $\overline{OP}$ is the **hypotenuse** of this triangle.

Icosahedron — a space figure with 20 faces

Intersection of Lines — the point at which two lines meet

Lines $\overline{AB}$ and $\overline{CD}$ **intersect** at point Y.

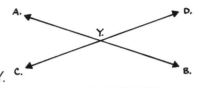

Intersection of Planes — a line formed by the set of points at which two planes meet

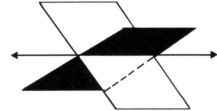

Lateral Faces — the plane surfaces of a space figure that are not bases

The **lateral faces** of this triangular prism are shaded.

Legs — sides adjacent to the right angle in a right triangle

$\overline{QP}$ and $\overline{QR}$ are **legs** in this triangle.

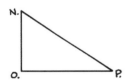

Line Segment — part of a line consisting of a path between two endpoints

$\overline{AB}$ and $\overline{CD}$ are line **segments**.

Linear Measure — (or length) the measure of distance between two points along a line

Liter — a metric system of measurement for liquid capacity

Measurement — the process of finding the length, area, capacity, or amount of something

Meter — a metric system unit of linear measurement

Metric System — a system of measurement based on the decimal system

Midpoint — a point that divides a line segment into two congruent segments

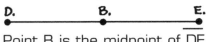

Point B is the midpoint of DE.

Nonagon — a nine-sided polygon

Octagon — an eight-sided polygon

Octahedron — a space figure with eight faces

Parallel Lines — lines in the same plane which do not intersect

These lines are **parallel**.

Parallelogram — a quadrilateral whose opposite sides are parallel

Pentagon — a five-sided polygon

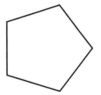

Perimeter — the distance around the outside of a closed figure

Perpendicular Lines — two lines in the same plane
that intersect at right angles

Lines XY and VW are
perpendicular to one another.

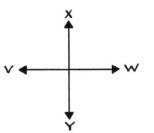

Pi — the ratio of a circle's circumference to its diameter.
The symbol π signifies pi.
pi = 3.14159265 (a non-terminating decimal)

Plane Figure — a set of points in the same plane enclosing a region

Figures A and B
are **plane figures**.

Polygon — a simple, closed plane figure having line segments as sides

Polyhedron — space figures formed by intersecting plane surfaces called faces

Prism — a space figure with two parallel, congruent polygonal faces called bases

(A prism is named by the shape of its bases.)

triangular prism   *rectangular prism*

Pyramid — a space figure having one polygon base and four triangular faces that have a common vertex

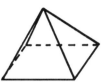

Pythagorean Theorem — a proposition stating that the sum of the squares of the two shorter sides of a right triangle is equal to the square of the third side

In triangle ABC, $AB^2 + BC^2 = CA^2$

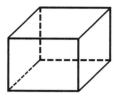

Quadrilateral — a four-sided polygon

Radius — a line segment having one endpoint in the center of a circle and another on the circle.

$\overline{FG}$ is the ***radius*** of this circle.

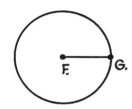

Ray — a portion of a line extending from one endpoint in one direction indefinitely

Rectangle — a parallelogram having four right angles

Rhombus — a parallelogram having congruent sides

Segment — two points and all of the points on the line or arc between them

Similarity — a property of geometric figures having angles of the same size

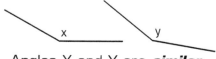

Angles X and Y are ***similar***.   Triangles A and B are ***similar***.

Skew Lines — lines that are not in the same plane and do not intersect

Slide — moving a figure without turning or flipping it
The shape or size of a figure
is not changed by a slide.

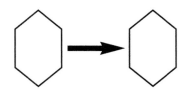

Space Figure — a figure which consists of a set of points in two or more planes

Sphere — a space figure formed by a set of points
equidistant from a center point

Square — rectangle with congruent sides

Surface — a region lying on one plane

Surface Area — the space covered by a plane region or by the faces of a space figure

Symmetric Figure — a figure having two halves that are reflections of one another;
a line of symmetry divides the figure into two congruent parts

These figures are **symmetric**. ST is a line of **symmetry**.

Tangent — a line which touches a curve at only one point

Line HG is **tangent** to the circle at point X

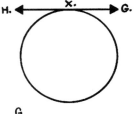

Transversal — a line that intersects
two or more parallel lines

$\overline{GH}$ is a **transversal** of lines AB and CD.

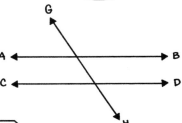

Trapezoid — a quadrilateral having only
two sides that are parallel

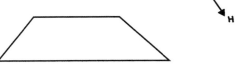

Triangle — a three-sided polygon

Acute Triangle — a triangle in which all three angles are less than 90°

Equilateral Triangle — a triangle with three congruent sides and three congruent angles

Isosceles Triangle — a triangle with at least two congruent sides

Obtuse Triangle — a triangle having one angle greater than 90°

Right Triangle — a triangle having one 90° angle

Scalene Triangle — a triangle in which no two sides are congruent

Turn — a move in geometry which involves turning, but not flipping, a figure

The size or shape of a figure is not changed by a turn.

Unit — 1. the first whole number
2. a determined quantity used as a standard for measurement

Vertex — a common endpoint of two rays forming an angle, of two line segments forming sides of a polygon, or of two planes forming a polyhedron

Point Z is the **vertex** of this angle.

Vertical — a line that is perpendicular to a horizontal base line

Line KL is **vertical**.

K

L

Volume — the measure of capacity or space enclosed by a space figure

Average — the sum of a set of numbers divided by the number of addends
The average of   1, 2, 7, 3, 8, and 9 =
$$\frac{1 + 2 + 7 + 3 + 8 + 9}{6} = 5$$

Chance — the probability or likelihood of an occurrence

Combination — a selection of a set of things from a larger set without regard to order

Coordinate Plane — a grid on a plane with two perpendicular lines of axes

Coordinates — a pair of numbers which give the location of a point on a plane

Counting Principle — a way to find the number of possible outcomes of an event with multiple stages. The total number of possible outcomes is the product of the outcomes of each stage.

Data — figures, facts, or information

Dependent Events — two events in which the result of the first event affects the outcome of the second event

Event — a set of one or more outcomes

Frequency — the number of times a given item occurs in a set of data

Frequency Graph — a way to organize and picture data using a grid

Frequency Table — data arranged on a table to show how often events occur

Function — a set of ordered pairs of numbers that follow a function rule and in which no two first numbers are the same
{(2, 5) (3, 6) (4, 7) (5, 8) (6, 9)}
The rule for this set is to add one.

Graph — a drawing showing relationships between sets of numbers

Bar Graph — a graph that represents data with bars

Circle Graph — a graph that represents data by showing a circle divided into segments

Line Graph — a graph that uses lines to show changes in data over time

Grid — a set of horizontal and vertical lines spaced uniformly

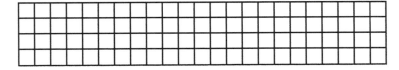

Histogram — a bar graph showing frequency data

Independent Events — events whose outcomes have no effect on later events

Interval — amount of space or time

Mean — average; the sum of numbers in a set divided by the number of addends
The **mean** of {6, 8, 9, 19, and 38} is $\frac{80}{5}$ or 16.

Median — the middle number in a set of numbers. The median is determined by arranging numbers in order from lowest to highest and by counting to the middle
The **median** of {3, 8, 12, 17, 20, 23, and 27} is 17.

Mode — the score or number found most frequently in a set of numbers
The **mode** of {36, 3, 12, 9, 7, 9, 23, 4, 12, 7} is 12.

Odds — the numerical likelihood of a chosen outcome in comparison to another

The odds against being eaten by a cat are higher than the odds in favor.

Odds Against — the numerical chance that an outcome will not be chosen; the ratio of unfavorable outcomes to favorable outcomes

Odds In Favor — the numerical chance that an outcome will be chosen; the ratio of favorable outcomes to unfavorable outcomes

Outcome — a possible result in a probability experiment

Permutation — an arrangement of data in a definite order

Pictograph — a graph that uses pictures or symbols to represent numbers

Prediction — the projection into the future of possible outcomes, based on data at hand

Probability — a study of the likelihood that an event will occur

Random — an experiment in which the results are not predictable, even when repeated

Range — the difference between the greatest number and the least number in a collection of data

Sampling — a method of gaining data from a selection of a larger amount of data, in order to make predictions about larger amounts

Statistics — numerical observations or data

Absolute Value — the distance of a number from zero on a number line

Axes — the two perpendicular number lines in a coordinate plane that intersect at 0

Coefficient — the number value in a mathematical expression
In the expression **8x**, **8** is the coefficient of **x**.

Coordinate Plane — a grid on a plane with two perpendicular lines of axes

Coordinates — a pair of numbers that give the location of a point on a plane

Coincide — the intersection of two lines in more than one point

Collinear Points — points that lie on the same line

Coordinate — a number paired to a point

Coordinates — a pair of numbers paired with a point

Coordinate Plane — a grid on a plane with two perpendicular number lines (axes)

Cube Numeration — a number raised to the third power (8^3)

Equation — a mathematical sentence which states that two expressions are equal
$7 \times 9 + 3 + (4 \times 5) = 86$

Equivalent Equations — equations that have the same solution

Evaluate — to substitute a number for each variable in an expression and simplify the expression

Function — a set of ordered pairs (x, y) where for each value of x, there is only one value of y

Inequality — a number sentence showing that two numerals or two groups of numerals stand for different amounts or numbers
The signs < *(is less than)*, > *(is greater than)*, and ≠ *(is not equal to)* show inequality.
$7 + 5 < 17 - 3$

Integers — the set of numbers greater than and less than zero

negative integers — the set of integers less than zero

positive integers — the set of integers greater than zero

231

Linear Equation — an equation whose graph is a straight line

Open Sentence — a number sentence with a variable

Opposites — two numbers on a number line that are the same distance from 0 on each side

Opposite Property — a property that states that if the sum of two numbers is 0, then each number is the opposite of the other
$-4 + 4 = 0$; -4 and 4 are **opposites**

Ordered Pair — a pair of numbers in a certain order with the order being of significance

A different kind of radical sign.

Radical Sign — the square root symbol $\sqrt{\quad}$

Rational Numbers — a number that can be written as the quotient of two numbers (A terminating or repeating decimal is rational.)

Real Numbers — any number that is a positive number, a negative number, or 0

Reciprocals — two numbers whose product is one

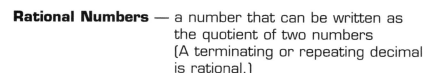
$\frac{1}{3}$ and $\frac{3}{1}$ are **reciprocals** because $\frac{1}{3} \times \frac{3}{1} = 1$.

Replacement Set — a set of numbers that could replace a variable in a number sentence

Solution Set — the set of possible solutions for a number sentence

Square Root — a number that yields a given product when multiplied by itself
The **square root** of 25 is 5 because $5 \times 5 = 25$

Scientific Notation — a number expressed as a decimal number (usually with an absolute value less than 10) multiplied by a power of 10
$4.53 \times 10^3 = 4530$

Solution — the number that replaces a variable to complete an equation

Variable — a symbol in a number sentence that could be replaced by a number
In **$3 + 9x = 903$**, **x** is the variable.

X-Axis — the horizontal number line on a coordinate grid

Y-Axis — the vertical number line on a coordinate grid

Accuracy — the correctness of all parts of a problem or solution

Common Element — numbers or units that are the same kind
In the problem below, 12.5 and 1.6 are common elements (decimals);
Also, $\frac{1}{2}$, 80, and 1.6 are common elements (apples)

12.5 apples + $\frac{1}{2}$ apples + 52 coconuts + 80 apples + 1.6 apples

Communication — the use of pictures, numbers, symbols, diagram, and/or words to show clearly the process used to solve a problem

Conceptual Understanding — comprehension of what a problem is and how that problem can be translated into mathematical numbers, symbols, or ideas

Data — facts, figures, or information containing numbers

Diagram — a drawing or sketch used to show data

Equation — a mathematical sentence which states that two expressions are equal (The equation often includes one or more variables.)
$$35x + 5y = 900$$

Estimation — the process of arriving at an approximate answer to a problem

Formula — a general fact, rule, principle, or process, expressed in mathematical symbols

Graph — a drawing showing relationships between sets of numbers

Guess & Check — a problem-solving strategy that involves making an estimate, then counting or calculating to see if the estimate is correct

Logic — a way of solving problems using principles of reasoning

Mental Math — solving problems in one's mind without the use of paper, pencil, calculator, computer, or any other external aid

Model — a three-dimensional representation of an item or event

Number Line — a line that has numbers corresponding to points along it

-4   -3   -2   -1   0   1   2   3   4

233

Open-Ended Problem — a problem that has more than one solution

Operations — processes applied to numbers, usually addition, subtraction, multiplication, and division

Pattern — an orderly arrangement of things or facts, often involving repetition

Problem — a question to be answered

Proportion — a number statement of equality between two ratios (A proportion may have a variable in it.)

$$\frac{12}{20} = \frac{x}{144}$$

Reasonableness — sensibleness of an answer

Simplify — reduce or change a problem, expression, or equation to less complicated terms by combining like terms

Solution — the answer to a problem; a number that replaces a variable to complete an equation

Solution Set — a set of possible solutions for a number problem or equation

Symbol — a number, picture, or letter that stands for a quantity

Table — a chart or grid that is a way to organize data into rows and columns

Trial & Error — a problem-solving strategy that involves choosing different solutions and testing them

Variable — a letter, symbol, or number that could be replaced by a number

Verification — process of defending or proving a solution

INDEX

Better Grades & Higher Test Scores / MATH
©Incentive Publications, Inc., Nashville, TN

G
geometry, 122-135, 137-146
  angles, 123, 124-127
  area, 140-`41
  circles, 131
  capacity, 146
  circumference, 139
  congruent figures, 132
  lines, 122-123, 126
  line segments, 123, 127
  perimeter, 138-139
  plane figures, 122-133
  planes, 122-123
  points, 122
  polygons, 128-130, 140-141
  quadrilaterals, 130
  similar figures, 132
  space figures, 134-135
  surface area, 142-143
  symmetry, 133
  tessellations, 133
  transformations, 133
  triangles, 129
  volume, 144-145
get healthy, 24
get organized, 18-23
Goldbach's Guess, 91
grain, 151
graphs, 156-162, 189-190
  bar, 156-157
  circle, 159
  coordinate, 189
  frequency, 156
  histogram, 158
  line, 161
  pictograph, 160
  problem-solving, 200
  scattergram, 162
greatest common factors, 66
guess & check strategy, 196

H
hand, 152
health, 23
histograms, 158
horsepower, 152

I
identity property, 32, 89
improper fractions, 96
inequalities, 187
infinity, 69
independent events, 166
integers, 31, 174
  addition, 176
  comparing, 175
  division, 177
  multiplication, 177
  negative, 31, 174
  operations with, 176-177
  ordering, 175
  positive, 31, 174
  subtraction, 176
interest, 120
intersecting lines, 122
intersecting planes, 123
intersection of sets, 57

International Date Line, 147, 149
irrational numbers, 72
isosceles triangle, 129

J
joule, 152

K
knot, 151

L
latitude, 147
least common denominator, 101
least common multiples, 67, 101
length, 34, 137
light year, 151
like fractions, 101
like terms, 179
line segments, 123, 127
line graphs, 161
lines, 122-123, 126-127
  intersecting, 122
  parallel, 122
  perpendicular, 122
  rays, 123
  segments, 123, 127
  transversal, 126
listening skills, 46
logic, 204
longitude, 147
lowest terms, 100

M
math tools, 30-37
mathematical expressions, 178
  coefficient, 178
  evaluating, 179
  simplifying, 179
  terms, 178
  variable, 178
mathematical symbols, 30
mean, 155
measurement, 34-37, 136-152
  area, 34, 140-141
  capacity, 35, 146
  circumference, 138
  conversions, 37
  dry, 146
  Earth, 147-149
  English System, 34. 37
  latitude, 150
  length, 34, 137
  liquid, 146
  longitude, 150
  metric, 34, 37, 136, 137, 146
  perimeter, 138
  plane figures, 137-141
  rate, 149
  space figures, 142-145
  surface area, 142-143
  temperature, 36, 150
  time, 36, 148-149
  units, 34
  U.S. customary units, 34, 37, 136, 137, 146
  volume, 35, 144-145
  weight, 36, 136
median, 155
mental math, 203

Better Grades & Higher Test Scores / MATH
©Incentive Publications, Inc., Nashville, TN

Better Grades & Higher Test Scores / MATH
©Incentive Publications, Inc., Nashville, TN